INTERMEDIATE

ALGEBRA

The Why and the How

First Edition

Mathew Baxter
Florida Gulf Coast University

cognella® | ACADEMIC PUBLISHING

Bassim Hamadeh, CEO and Publisher
Jess Estrella, Senior Graphic Designer
Jennifer McCarthy, Acquisitions Editor
Gem Rabanera, Project Editor
Abbey Hastings, Associate Production Editor
Chris Snipes, Interior Designer
Natalie Piccotti, Senior Marketing Manager
Kassie Graves, Director of Acquisitions
Jamie Giganti, Senior Managing Editor

Printed in the United States of America

ISBN: 978-1-5165-0303-2 (pbk) / 978-1-5165-0304-9 (br)

cognella® | ACADEMIC PUBLISHING

CONTENTS

CONTENTS

CHAPTER

1

The Real Number System

1.1: Real Numbers and Their Properties

We begin our studies with the definitions of the kinds of numbers we will be using in this text: real numbers. We also discuss the important properties of real numbers.

Real Numbers

First, we'll discuss sets in general, so that we have good terminology when we move on to sets of numbers.

Sets are groups of objects, called elements. When we list the elements of a set, we usually make a comma-separated list surrounded by braces.

Ex) The set of days has seven elements: the five weekdays as well as Saturday and Sunday. If we call the set of days D for short, then

$$D = \{\text{Monday, Tuesday, Wednesday, Thursday, Friday, Saturday, Sunday}\}.$$

Ex) When we flip a fair coin, the set of possibilities consists of two elements: heads and tails. In statistics, this set of outcomes is called the **sample space**.

We need to be able to tell if an element actually belongs to a set. We use the raised Greek letter *epsilon*, ε, to mean "is an element of."

Ex) If we call the set of days D, then Monday is an element of the set D. We can write this as Monday ε D.

Ex) In our coin example, if we call the outcome of heads H and we call the set of all outcomes when flipping a fair coin Ω (a capital omega), we can write $H \in \Omega$. In this example, Ω consists of "heads" or "tails," H or T for short. So we can say $\Omega = \{H,T\}$.

Sets can be infinite. We use an ellipsis (a trail of periods, like "...") to indicate that a set continues. For example, the set of positive even numbers {2, 4, 6, 8, ...} is an infinite set.

In this text, the most important sets will be sets of numbers. See Figure 1-1.

The natural numbers are the counting numbers 1, 2, 3, … .

The natural numbers are denoted by $\mathbb{N}$. So we may write $\mathbb{N} = \{1, 2, 3, \ldots\}$.

The whole numbers are the natural numbers, including 0. So the whole numbers are 0, 1, 2, … . The set of whole numbers will not be used much in this text. In Figure 1-1, whole numbers would be between natural numbers and integers.

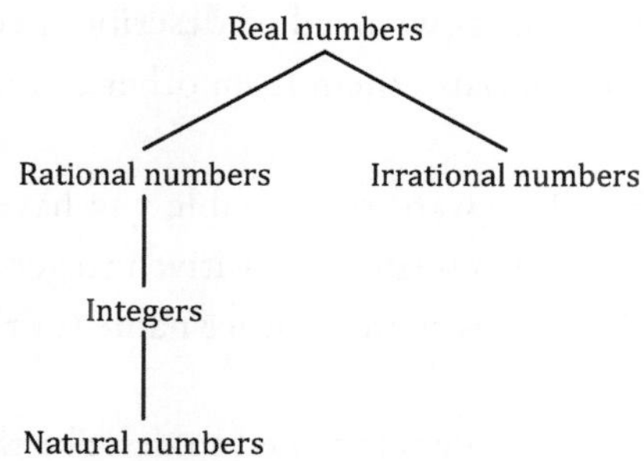

Fig. 1-1
Types of Numbers

The integers are the natural numbers, their opposites, and zero.

Recall that the opposite of a number is the same number across from zero on a number line. So the opposite of 2 is −2, and the opposite of −3 is 3. See Figure 1-2.

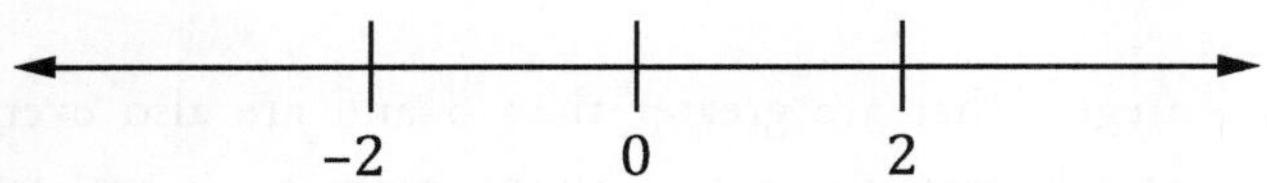

Fig. 1-2
2 is the opposite of −2

The integers are denoted by $\mathbb{Z}$. The "Z" stands for *zahlen*, which is the German word for *integer*. So we can write $\mathbb{Z} = \{\ldots, -2, -1, 0, 1, 2, \ldots\}$.

Thus, the integers are all of the usual tick marks when we make a number line. See Figure 1-3.

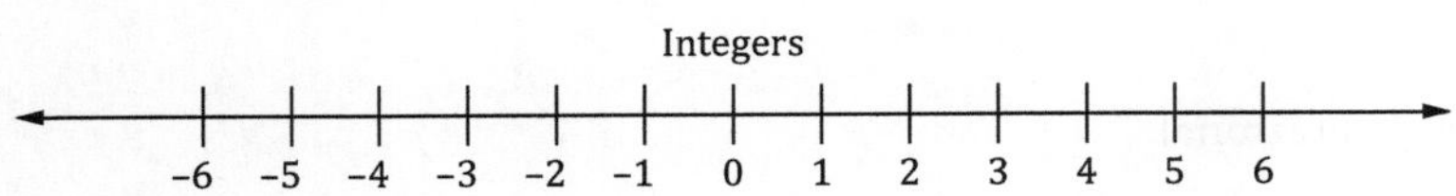

Fig. 1-3
Integers at the lattice points

Ex) The number -7 is an integer, and the number 14 is an integer. But of the two of them, only 14 is a natural number because it is positive.

Definition: The rational numbers (or rationals) are the set of all numbers $\frac{a}{b}$ where a and b are both integers and b cannot be zero.

Sometimes we need to describe specific elements of a set. These elements have a property that differentiates them from other elements and are identified using a colon or a vertical bar.

Ex) If we want the variable x to have the property that x is an integer, then we write $x \in \mathbb{Z}$. If we want to describe all positive integers, then we would write $\{y \in \mathbb{Z} : y > 0\}$ or $\{y \in \mathbb{Z} | y > 0\}$. Think about it: is there another name for this set?

Ex) Suppose we wanted to describe the integers greater than 5. We can write this as $\{a \in \mathbb{Z} : a > 5\}$. First, we are describing a general statement, saying that we want our arbitrary number a to be an integer. Then, after the "such that" colon, we are describing a more specific trait that we want a to have.

YOU TRY IT: Using set notation, describe all of the integers less than -2.

Ex) Let's describe the integers that are greater than 5 and are also even. We can separate the requirements on our arbitrary integer a using commas, so we could write this set as $\{a \in \mathbb{Z} : a > 5, a \text{ even}\}$.

The rationals are denoted by $\mathbb{Q}$. Look again at the definition and you will see why the letter "Q" was chosen. It stands for "quotients" (of integers). We can write the definition of the rational numbers in set notation as

$$\mathbb{Q} = \left\{ \frac{a}{b} : a \in \mathbb{Z}, b \in \mathbb{Z}, b \neq 0 \right\}.$$

Ex) The number $\frac{4}{7}$ is rational.

Ex) Anything below the rational numbers in Figure 1-1 is also a rational number. For example, −4 is a rational number, because we can write it as the quotient of −4 and 1: $\frac{-4}{1}$. And 18 is a rational number, because $18 = \frac{18}{1}$.

Ex) The number 0.333 ... is rational, because this is the decimal representation of $\frac{1}{3}$. The number $0.1\overline{6}$ is rational, because this is the decimal representation of $\frac{1}{6}$.

What's important to remember about the previous example is that *repeating decimals can be represented as fractions*, so they are rational numbers. This is one key distinction between rational numbers and irrational numbers.

Any real number that is not rational is *irrational.* For example, $\sqrt{2}$ is irrational because it cannot be written as the quotient of integers. Another example is π, because $\pi = 3.141592653\ldots$ has a *nonrepeating nonterminating* decimal representation.

To clarify:

Number Type	Decimal representation	Written as a fraction?
Rational	Repeating or terminating	Can be written as a fraction
Irrational	Nonrepeating and nonterminating	Cannot be written as a fraction

Note that the decimal representation of a number must be *both* nonrepeating and nonterminating for that number to be irrational.

Ex) The decimal 3.141 is nonrepeating but it terminates, so it is a rational number. We can write it as $\frac{3{,}141}{1{,}000}$.

Ex) The decimal −0.8888 ... is nonterminating but it repeats, so it is a rational number. This is $-\frac{8}{9}$.

One final thing worth mentioning about the rationals is the restriction that the denominator may not be zero. Division by zero is not allowed. You cannot divide something into zero parts.

The real numbers make up the set that contains both the rational numbers and the irrational numbers. The notation for the real numbers is $\mathbb{R}$.

Example: Use the classifications of *integer, real number, rational number, irrational number,* and/or *natural number* to describe each of these numbers:

1. $\frac{3}{2}$
2. v–73
3. $\frac{5}{6} = 0.8333\ldots$
4. $\sqrt{7} = 2.64575\ldots$

Solution:

1. $\frac{3}{2}$ is a quotient of the integers 3 and 2, which makes it a rational number. Remember that it is also anything higher than the rationals in Figure 1-1, so it is also a real number.
2. -73 is an integer (since it is negative, it is not a natural number). Because it is an integer, it is also a rational number and a real number.
3. Note that $\frac{5}{6}$ has a repeating decimal representation, so it is a rational number. It is also a real number.
4. The difference here is that $\sqrt{7}$ has a decimal representation that is nonrepeating and nonterminating, so it is an irrational number. It is also a real number.

Properties of Real Numbers

When we add two positive real numbers together, the result is another positive real number. For example, 7 and 12 are both positive numbers, and their sum, 19, is another positive real number. If we subtract two positive real numbers, the result may not be positive. We know that $12 - 7 = 5$, and 5 is positive, but if we subtract 12 from 7 we get $7 - 12 = -5$. The real numbers are imbued with an ordering that lets us compare any two of them and decide which is larger. So when we subtract, we have to make a note about which number is larger.

If we do

larger number – smaller number ... the result is positive,

and if we do

smaller number – larger number ... the result is negative.

Ex) If we subtract 132 from 133, we end up with a positive number, because 133 is larger than 132. So $133 - 132 = 1$.

Ex) If we subtract 7 from 5, the result is a negative number, because we are subtracting a larger number from a smaller number, which is case two above. So $5 - 7 = -2$.

We can add numbers in any order, but we cannot subtract numbers in any order. We can summarize this by saying that addition of real numbers is commutative, whereas subtraction is not. We have that $a + b = b + a$, but $a - b$ is not necessarily equal to $b - a$.

Ex) $3 + 4$ and $4 + 3$ both equal 7. But $4 - 3 = 1$ while $3 - 4 = -1$.

Multiplication and division present a similar situation. We can multiply two numbers in any order, and the result is the same. But we cannot divide numbers in any order.

Ex) $3 \cdot 4 = 12$, and this is the same as $4 \cdot 3 = 12$. But note that $3 \div 4$ and $4 \div 3$ are different numbers (0.75 and $1.\overline{3}$, respectively).

Another nice property of the real numbers is that addition (and multiplication) is associative, meaning that we can add multiple numbers in any order from left to right. If we add the numbers 3, 4, and 12 from left to right, we can add 3 and 4 first or we can add 4 and 12 first. If we use parentheses to denote this, we are saying that $(3+4)+12=3+(4+12)$, where we are doing the addition in the parentheses first.

The associative property can be extended to any amount of numbers we are adding (or multiplying) together.

Ex) To add the numbers −2, 7, 5, and 15 from left to right, we are allowed to do $(-2+7)+5+15$, or $-2+(7+5)+15$, or $-2+7+(5+15)$.

Another property of the real numbers is that there is an additive (and multiplicative) identity. So if you take any number, and add 0, you get the same number back. For example, $5+0=5$ and $0+5=5$. There is nothing special about the 5 here. It is the 0 that we call the additive identity.

Similarly, we can multiply a number by 1 to get the same number back. For example, $8\cdot 1=1\cdot 8=8$. Since we can do this for any number, not just 8, we call 1 the multiplicative identity.

Two other properties to discuss are the inverse properties, which will be extremely important when we are solving equations in the next chapter. Given any number, we would like to add something to it so we get back to the additive identity, 0. This "something" that we are adding is the additive inverse.

Ex) The additive inverse of 12 is -12, because when we add these two numbers together we get the additive identity: $12+(-12)=0$, or $-12+12=0$.

Ex) The additive inverse of -2.5 is 2.5, because $-2.5+2.5=0$.

So to find the additive inverse of a number, give its opposite.

Similarly for multiplication, the multiplicative inverse of a is the number you multiply by a to get back to the multiplicative identity 1.

Ex) The multiplicative inverse of 7 is $\frac{1}{7}$, because $7\cdot\frac{1}{7}=\frac{7}{1}\cdot\frac{1}{7}=1$.

Ex) The multiplicative inverse of $\frac{3}{4}$ is $\frac{4}{3}$ because $\frac{3}{4}\cdot\frac{4}{3}=1$.

So to find the multiplicative inverse of a number, give its reciprocal.

YOU TRY IT 2: Find the sum of the additive inverse of 8 and the multiplicative inverse of 8.

The final property to discuss here is the only one that combines the operations of addition and multiplication. It is the distributive property, which tells us that $a(b+c)=ab+ac$. This is the property that allows us to distribute, where we have

$$2(x+y)=2x+2y,$$

but it also allows us to factor, where we can write

$$2x + 2y = 2(x + y).$$

The table below summarizes the properties of the real numbers.

YOU TRY IT ANSWERS: 1) $\{x \in \mathbb{Z} : x < -2\}$ 2) $\frac{65}{8}$

Property Name	Property Example ($a, b, c \in \mathbb{R}$)
Commutativity of Addition	$a + b = b + a$
Commutativity of Multiplication	$a \cdot b = b \cdot a$
Associativity of Addition	$a + (b + c) = (a + b) + c$
Associativity of Multiplication	$(a \cdot b) \cdot c = a \cdot (b \cdot c)$
Additive Identity	$a + 0 = 0 + a = a$
Multiplicative Identity	$a \cdot 1 = 1 \cdot a = a$
Additive Inverse	$a + (-a) = -a + a = 0$
Multiplicative Inverse	$a \cdot \frac{1}{a} = \frac{1}{a} \cdot a = 1, a \neq 0$
Distributivity	$a(b + c) = ab + ac$

Section 1.1 Exercises

1. What is the difference between a rational number and an irrational number?
2. Can a number be *both* rational *and* irrational? If so, give an example. If not, explain why.

Classify the numbers in problems 3–9 as real, rational, irrational, integer, and/or natural.

3. -2
4. $4.\overline{6}$
5. 5
6. $\sqrt{3}$
7. $\sqrt{4}$

8. $\frac{7}{2}$
9. $\frac{\sqrt{7}}{2}$
10. a. Explain why $0.14159265\ldots = \pi - 3$ is an irrational number.
    b. Explain why $0.12345678987654321234567898765432l\ldots$ is a rational number. Its decimal representation is the digits 1 through 9 in ascending order, then descending back to 1, and so on.
11. Give the ...
    a. Opposite of 7
    b. Reciprocal of 7
    c. Opposite of $\frac{3}{7}$
    d. Reciprocal of $\frac{3}{7}$
12. Addition of real numbers is commutative, but in general subtraction of real numbers is not. For example, $4-7 \neq 7-4$. What must be true of a and b so that $a-b=b-a$?
13. Give an example to show why subtraction is not associative. That is, find numbers a, b, and c, and calculate $a-(b-c)$ and $(a-b)-c$.
14. Which of the properties of real numbers allows us to write that $48+96=24(2+4)$?
15. Which of the properties of real numbers allows us to FOIL? *FOIL* stands for "first, outer, inner, last" and it's how we multiply out $(a+b)(c+d)$. We get $(a+b)(c+d)=a(c+d)+b(c+d)=ac+ad+bc+bd$.
16. Which of the properties of real numbers allows us to add four numbers from left to right in any order we want?
17. Which of the properties of real numbers tells us that if we're multiplying -37 and $\frac{67}{11}$, then we can write down either number first?
18. Is division commutative for all real numbers (except zero)? Explain why or why not.
19. For a property to be true, it needs to hold for all real numbers.
    a. Suppose that division has an associative property. This means we claim that

$$a \div (b \div c) = (a \div b) \div c.$$

    Calculate both sides of this equation when $a=2$, $b=1$, and $c=1$. Does this mean the property holds for all real numbers?

b. Find different numbers for a, b, and c, calculate both sides of the so-called associative property of addition, and show that this property does not hold for all real numbers.

20. Looking ahead: Write down the sample space that describes the outcomes when a fair 6-sided die is rolled. Meaning, write down a set Ω that lists the possibilities when rolling the die.

1.2: The Four Basic Operations

In this section we discuss the four basic operations on real numbers: addition, subtraction, multiplication, and division. We also talk about another operation: absolute value.

For the most part, we will discuss these operations in the context of rational numbers. (We will discuss how these operations act on irrational numbers in detail in Chapter 7.)

In the last part of this section, we will discuss how the four operations act together in mathematical expressions.

Addition and Subtraction

When adding two positive integers, the result is positive.

Ex) $37 + 750 = 787$

One convention that we adopt is that if a and b are real numbers, then

$$a + (-b) = a - b.$$

So adding the opposite of a number is really like subtraction. Therefore, when we add two negative integers, the result is negative.

Ex) $-3 + (-7) = -3 - 7 = -10$

When adding one positive and one negative integer, follow the rules discussed in the previous section, remembering that adding a negative is like subtracting.

Ex) $-3+7=4$. The result is positive, because 7 is larger than 3.

Ex) $6+(-15)$ is the same as $6-15=-9$. The result is negative because 15 is larger than 6.

When adding or subtracting negative numbers, we have to remember that subtracting a negative number is like adding a positive number: We are reversing an opposite.

Ex) If we are solving for $-7-(-12)$, we are taking away the opposite of 12. Think of the opposite of taking away as giving back, so we are really adding. So $-7-(-12)=-7+12=5$.

Ex) Suppose you have \$10, and your friend takes away the opposite of \$5. This is like your friend giving you \$5. In mathematical notation, you have $10-(-5)=15$ dollars.

Below is a summary of what we have discussed.

When adding ...	**the result is ...**
two positive numbers	positive
one positive number and one negative number	like computing $a-b$ where $a, b > 0$; the result is positive if $a > b$ and negative if $b > a$
two negative numbers	negative

When adding and subtracting rational numbers, it is important to have a common denominator.

Ex) If one person eats one and one-half pies, and another eats one-sixth of a pie, how much pie is eaten? Well, consider that one and one-half is $1\frac{1}{2}$. Writing this as a complex fraction, we have $1+\frac{1}{2}=\frac{2}{2}+\frac{1}{2}=\frac{3}{2}$. Then, adding, we have $\frac{3}{2}+\frac{1}{6}=\frac{3}{2}\cdot\frac{3}{3}+\frac{1}{6}=\frac{9}{6}+\frac{1}{6}=\frac{10}{6}=\frac{5}{3}$. So in total, five-thirds pies are eaten.

Ex) $4-\frac{7}{2}=\frac{4}{1}\cdot\frac{2}{2}-\frac{7}{2}=\frac{8}{2}-\frac{7}{2}=\frac{1}{2}$

Multiplication and Division

When multiplying a times b, think of having a copies of b. So if we have $2 \cdot 3$, then we have 2 sets of 3 objects, or a total of 6 objects. This also works well if you are multiplying one positive number and one negative number.

Ex) $-3 \cdot 2 = -6$, because we have 2 copies of -3, or -6 in total.

Ex) $4(-6)$ is like 4 copies of -6, so there is a total of -24.

How does this make sense when we multiply two negative numbers? Why does $(-2)(-3) = 6$? The answer goes back to reversing opposites. If "taking away" is a negative, then the reverse of "taking away" is actually "giving," or a positive.

Ex) Suppose you pay rent out of your bank account. Think of $2(-700)$ as you making two rent payments of \$700 each, so you are taking away \$1400 from your account. Then what does $(-2)(-700)$ represent? It is the opposite of deducting two rent payments from your account, so you are actually adding \$1400 to your account. If only it were that easy.

When multiplying rational numbers, we take the product of the numerators divided by the product of the denominators.

Ex) If we multiply $\frac{2}{3}$ times $\frac{3}{2}$, we have $\frac{2}{3} \cdot \frac{3}{2} = \frac{6}{6} = 1$.

Ex) If we multiply 6 times $\frac{3}{5}$, we have $\frac{6}{1} \cdot \frac{3}{5} = \frac{6 \cdot 3}{1 \cdot 5} = \frac{18}{5}$.

Ex) $-7\left(\frac{3}{14}\right) = -\frac{7}{1} \cdot \frac{3}{14} = -\frac{3}{2}$

The same rules apply to division. If only one of the numbers that we are dividing is negative, the result is negative. However, if both numbers are negative, the result is positive.

Ex) $-4 \div 2 = -2$

Ex) $-4 \div (-2) = 2$, since we are dividing two negative numbers.

Below is a summary of what we've discussed.

When multiplying or dividing ...	the result is ...
two positive numbers	Positive
one positive and one negative number	Negative
two negative numbers	Positive

When dividing fractions, remember that dividing by a number is the same as multiplying by the reciprocal of that number. So the second fraction is flipped.

Ex) $\frac{2}{3} \div \frac{3}{2} = \frac{2}{3} \cdot \frac{2}{3} = \frac{2 \cdot 2}{3 \cdot 3} = \frac{4}{9}$

Ex) $\frac{2}{3} \div \left(-\frac{5}{3}\right) = \frac{2}{3} \cdot \left(-\frac{3}{5}\right) = \frac{2(-3)}{3(5)} = -\frac{2}{5}$

Ex) Dividing $-\frac{1}{3}$ by $-\frac{8}{9}$ gives us

$$\begin{aligned} -\frac{1}{3} \div \left(-\frac{8}{9}\right) &= -\frac{1}{3} \cdot \left(-\frac{9}{8}\right) \\ &= \frac{1 \cdot 9}{3 \cdot 8} \\ &= \frac{3}{8}. \end{aligned}$$

A helpful tip: When you're multiplying fractions like this, simplify as you go. For example, in the second-to-last step above, you want to divide the 3 into the 9 to get the answer of $\frac{3}{8}$, rather than multiplying straight across to get $\frac{9}{24}$ and having to simplify now.

At this point, it is important to note what negatives or opposites do to fractions. If you are dividing 4 by -1, the result is -4. But there are a few ways to write "-1 divided by 4." The first way is to have the negative sign sitting in the numerator, as in $\frac{-1}{4}$. The second way is to move the negative out front, as in $-\frac{1}{4}$ (this is the way this text will write negative fractions most often). The third way is to have the negative sign move to the denominator, as in $\frac{1}{-4}$ (this way is the least common, but all three are equally correct). Again, we have

$$\frac{-1}{4} = -\frac{1}{4} = \frac{1}{-4}.$$

We'll refer to this again when we talk about rational expressions in Section 6.2.

Absolute Value

On a number line, the starting point zero is sometimes called the origin. The absolute value of a number is the distance from the number to the origin. The notation for absolute value is to put vertical bars around the number. So the absolute value of a number x is written as $|x|$.

Ex) Since the number 5 is five units away from zero, we have $|5| = 5$. See Figure 1-4.

Ex) Since the number -5 is five units away from zero, we have $|-5| = 5$. See Figure 1-4.

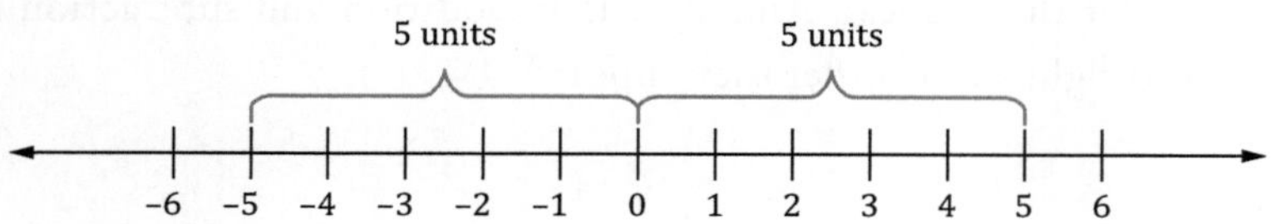

Fig. 1-4
$|5| = 5$ and $|-5| = 5$

We can summarize this by saying that if x is positive, then $|x| = x$, but if x is negative, then $|x| = -x$.

Ex) In $|-16|$, we are taking the absolute value of a negative number. So the answer is the opposite of this number, or $-(-16) = 16$.

Ex) $\left|-\frac{4}{7}\right| = \frac{4}{7}$

It is helpful to use absolute value when subtracting numbers. When you're subtracting numbers and trying to determine if the result is positive or negative, you keep the sign of the number that has the highest absolute value.

Ex) In the equation $4 - \frac{9}{2}$, since $\frac{9}{2}$ has a higher absolute value than 4, we know the answer will be negative. The result is $4 - \frac{9}{2} = \frac{8}{2} - \frac{9}{2} = -\frac{1}{2}$.

The Four Operations Together

If we look at a mathematical expression written out, like $1 + 1 \div 4$, it might seem that we are to add 1 and 1 together first. But if we are asked to add one and one-fourth, then we could write this out mathematically as $1+\frac{1}{4}$. It is important for us to be consistent with notation, so we need to make sure that $1 + 1 \div 4$ and $1+\frac{1}{4}$ have the same numerical value: $\frac{5}{4}$. This is why

multiplication and division are done from left to right before addition/subtraction.

So when we're simplifying mathematical expressions, multiplication and division are done from left to right in the order they appear. And after that, addition and subtraction operations are performed from left to right in the order they appear.

To the examples!

Example: Simplify $2 \div 1 \cdot 0$.

Solution: Solving from left to right, we have $2 \div 1 \cdot 0 = 2 \cdot 0 = 0$.

Example: Simplify $1-1+1-2\cdot 0$.

Solution: A good first step when simplifying is to scan the expression. Ignore the addition and subtraction for a moment, because the operation that is done first is the multiplication. It also helps to place parentheses around the two numbers that are affected by the operation, to make it clear what needs to happen. So here is how we should do this simplification:

$$\begin{aligned} 1-1+1-2\cdot 0 &= 1-1+1-(2\cdot 0) \\ &= 1-1+1-0 \\ &= (1-1)+1-0 \\ &= 0+1-0 \\ &= 1. \end{aligned}$$

Example: Simplify $2 \div 4 - 4 \div 2$.

Solution: A good way to think of this is $\frac{2}{4}-\frac{4}{2}$, which equals $\frac{1}{2}-2$. Then we can consider the solution as just a matter of subtracting fractions, in which case we have $\frac{1}{2}-2=\frac{1}{2}-\frac{4}{2}=-\frac{3}{2}$.

Example: Simplify $14+20\div 2\div 2+6\cdot 2-3$.

Solution: The first thing to take care of is the division. Then we move on to the other division and the multiplication. Again, it is a good idea to use parentheses to highlight what operation is happening next. This is how we should simplify:

$$\begin{aligned}
14+20\div 2\div 2+6\cdot 2-3 &= 14+(20\div 2)\div 2+6\cdot 2-3 \\
&= 14+10\div 2+6\cdot 2-3 \\
&= 14+(10\div 2)+6\cdot 2-3 \\
&= 14+5+(6\cdot 2)-3 \\
&= 14+5+12-3 \\
&= 19+12-3 \\
&= 31-3 \\
&= 28.
\end{aligned}$$

As you do more and more of these problems, you will need fewer and fewer steps. For example, you will not need to note the parentheses before you do every operation. But for now, adding parentheses is a great way to highlight the order of steps taken.

Example: Simplify $13+5-(-2)-5-(-6)+10$.

Solution: Since this expression contains only addition and subtraction, we move from left to right in successive steps:

$$\begin{aligned}
13+5-(-2)-5-(-6)+10 &= 18-(-2)-5-(-6)+10 \\
&= 18+2-5-(-6)+10 \\
&= 20-5-(-6)+10 \\
&= 15-(-6)+10 \\
&= 15+6+10 \\
&= 21+10 \\
&= 31.
\end{aligned}$$

Example: Simplify $64 \div 4 \cdot (-2) \div 8 \cdot (-3)$.

Solution: Again, since this expression contains only multiplication and division, we move from left to right in successive steps:

$$\begin{aligned} 64 \div 4 \cdot (-2) \div 8 \cdot (-3) &= 16 \cdot (-2) \div 8 \cdot (-3) \\ &= -32 \div 8 \cdot (-3) \\ &= -4 \cdot (-3) \\ &= 12. \end{aligned}$$

Example: Simplify $|-3| - |5-7| + \left|\frac{1}{2} - 3\right|$.

Solution: Simplifying inside the absolute values first, we find this:

$$\begin{aligned} |-3| - |5-7| + \left|\frac{1}{2} - 3\right| &= 3 - |-2| + \left|\frac{1}{2} - \frac{6}{2}\right| \\ &= 3 - 2 + \left|-\frac{5}{2}\right| \\ &= 1 + \frac{5}{2} \\ &= \frac{2}{2} + \frac{5}{2} \\ &= \frac{7}{2}. \end{aligned}$$

A last comment: it is worth noting that there are several ways to signify multiplication: using the dot, $\cdot$; using the "times" symbol, $\times$; using the asterisk, $*$; and using parentheses (parentheses are popular when we are dealing with negative numbers). To indicate division, we use a fraction bar or the division symbol, $\div$.

Example: Simplify $1 \cdot 2 \times 3 * 4 \div 6 \cdot \frac{1}{4}$.

Solution: From left to right, we have $2 \times 3 * 4 \div 6 \cdot \frac{1}{4} = 6 * 4 \div 6 \cdot \frac{1}{4} = 24 \div 6 \cdot \frac{1}{4} = 4 \cdot \frac{1}{4} = 1$.

Section 1.2 Exercises

In problems 1–38, simplify the expression.

1. $1+\frac{3}{7}$
2. $1+(-2)$
3. $\frac{3}{8}+\frac{2}{5}$
4. $4-8$
5. $47-56$
6. $-4-(-3)$
7. $\frac{2}{3}+\left(-\frac{2}{3}\right)$
8. $\frac{2}{3}-\left(-\frac{2}{3}\right)$
9. $4\cdot\frac{7}{2}$
10. $-2\cdot 5$
11. $-2\cdot(-5)$
12. $-2\div 1$
13. $-2\div(-1)$
14. $\frac{3}{23}\cdot\frac{46}{9}$
15. $\frac{3}{7}\cdot\frac{14}{27}$
16. $\frac{14}{15}\cdot\left(-\frac{5}{2}\right)$
17. $-\frac{14}{15}\cdot\left(-\frac{5}{2}\right)$
18. $\frac{14}{15}\div\frac{5}{2}$
19. $\frac{14}{15}\div\left(-\frac{5}{2}\right)$
20. $-\frac{14}{15}\div\left(-\frac{5}{2}\right)$
21. $|-3|$
22. $|3|$
23. $\left|\frac{3}{8}\right|$
24. $|0|$
25. $\left|-\frac{2}{9}\right|$

26. $|-5|+7\cdot 2-6$
27. $\left|-\frac{1}{3}\right|+\frac{5}{6}$
28. $4\cdot 5\div 2$
29. $3-6\cdot 5$
30. $18\div 2\div 3$
31. $2\cdot 3\cdot 4\cdot 5\cdot 0$
32. $4+7-3\cdot 10\div 2$
33. $3\cdot 4-8\cdot 2$
34. $18\div 2-15\div(-3)$
35. $3\div 7+4\div 7$
36. $14-16\div 4+2\cdot 6\div 3$
37. $64\div 2\div 2\div 2$
38. $2\div 2\div 2\div 2$

1.3: Order of Operations—PEMDAS

In this section we will discuss natural number exponents. After that, we will be prepared to handle the classical order of operations.

Exponents—A First Look

Exponents are a shorthand for multiplication. They are very helpful if we need to multiply a number by itself several times. We have the following definition:

$$x^n = x\cdot x\cdot \ldots \cdot x \ (n \text{ times}).$$

In words, we say that "x to the nth power is x multiplied by itself n times."

Ex) 3^4 is 4 copies of 3 multiplied out, so we have $3^4 = 3\cdot 3\cdot 3\cdot 3 = 9\cdot 3\cdot 3 = 27\cdot 3 = 81$.

Taking the second power and third power of a number has a direct reference to geometry. When we are raising a number to the second power, we are squaring it, so we write $6^2 = 36$, and we say

"6 squared equals 36." When we are raising a number to the third power, we are cubing it, so we write $2^3 = 8$ and we say "2 cubed equals 8."

Ex) $(-1)^5 = (-1)(-1)(-1)(-1)(-1) = 1(-1)(-1)(-1) = 1(-1) = -1$

When we raise fractions to powers, we put parentheses around them.

Ex) $\left(\frac{2}{5}\right)^3 = \frac{2}{5}\cdot\frac{2}{5}\cdot\frac{2}{5} = \frac{2\cdot 2\cdot 2}{5\cdot 5\cdot 5} = \frac{8}{25}$

Ex) $\left(\frac{3}{4}\right)^2 = \frac{3}{4}\cdot\frac{3}{4} = \frac{3\cdot 3}{4\cdot 4} = \frac{9}{16}$

When we raise a fraction to a power, we raise the numerator to that power and the denominator to that power:

$$\left(\frac{a}{b}\right)^n = \frac{a^n}{b^n}, \quad b \neq 0.$$

Ex) $\left(\frac{2}{3}\right)^2 = \frac{2^2}{3^2} = \frac{4}{9}$

Example: Simplify $\left(-\frac{1}{5}\right)^2$.

Solution: Recall our discussion about fractions and where we can place the negative sign. When squaring, move the negative to the top or the bottom number. The negative is definitely affected by the exponent:

$$\begin{aligned}\left(-\frac{1}{5}\right)^2 &= \left(\frac{-1}{5}\right)^2\\ &= \frac{(-1)^2}{5^2}\\ &= \frac{1}{25}.\end{aligned}$$

When the fraction has a low exponent like this one does, we could have just written out $\left(-\frac{1}{25}\right)\left(-\frac{1}{25}\right)=\frac{1}{25}$.

Example: Simplify $\left(-\frac{2}{3}\right)^5$.

Solution:

$$\begin{aligned}\left(-\frac{2}{3}\right)^5 &= \frac{(-2)^5}{3^5} \\ &= -\frac{32}{243}.\end{aligned}$$

YOU TRY IT: Evaluate $\left(-\frac{2}{5}\right)^3$.

PEMDAS

Let *P* stand for "parentheses," *E* stand for "exponents," *M* for "multiplication," *D* for "division," *A* for "addition," and *S* for "subtraction." We use the acronym PEMDAS, which stands for "Please Excuse My Dear Aunt Sally," to remember the order in which general expressions are simplified. We already covered the MDAS portion of PEMDAS in the last section. Multiplication and division are done from left to right, and then addition and subtraction are done last from left to right. But before doing any of these operations, we must take care of parentheses and exponents. Parentheses are to be worked from the inside out, and then exponents are resolved. Then we move on from there.

Example: Simplify $3+2^2\cdot 4$.

Solution: We must square 2 before doing the multiplication, so we have

$$\begin{aligned}3+2^2\cdot 4 &= 3+4\cdot 4 \\ &= 3+16 \\ &= 19.\end{aligned}$$

Example: Simplify $(-2)^2-2^2$.

Solution: Taking care of the parentheses first, we square −2 to get $4 - 2^2$. Then, we must take the exponent before we subtract, so we have $4 - 2^2 = 4 - 4 = 0$. A common mistake is made in evaluating -2^2; note that this is the opposite of 2^2, so it is −4.

Example: Simplify $-2(3-(2-5))$.

Solution: When there are multiple groups of parentheses, we work from the inside out, as follows:

$$\begin{aligned} -2(3-(2-5)) &= -2(3-(-3)) \\ &= -2(3+3) \\ &= -2(6) \\ &= -12. \end{aligned}$$

YOU TRY IT: Simplify $(-2)^2 + 2^2(-3)$.

Having a lot of parentheses can be confusing to the eye. So brackets and braces are used as well: [ ] and { }, respectively. These have the same priority of parentheses, and we work from the inside out.

Ex) $-2(-3(2+7(-3+2)))$ can be written as $-2\{-3[2+7+(-3+2)]\}$, which lets us clearly see what is "inside" (and therefore where to start simplifying).

Example: Simplify $-2\{-3[2+7+(-3+2)]\}$.

Solution: Start with the braces and work your way out:

$$\begin{aligned} -2\{-3[2+7+(-3+2)]\} &= -2\{-3[2+7(-1)]\} \\ &= -2\{-3[2-7]\} \\ &= -2\{-3[-5]\} \\ &= -2\{15\} \\ &= -30. \end{aligned}$$

Example: Simplify $4(-2)^2$.

Solution: A good habit is to say what this is by using words. It is "4 times negative 2 squared." The parentheses make clear that we must square -2 first, so it is even better to think of this as "4 times the square of -2." Thus, we have $4(-2)^2 = 4(4) = 16$.

When simplifying fractions, we can think of $\frac{1+3-2}{4+6\cdot 2}$, for example, as $(1+3-2) \div (4+6\cdot 2)$. Because of the parentheses, we can simplify the numerator and the denominator separately.

Ex) $\frac{1+3-2}{4+6\cdot 2} = \frac{4-2}{4+12} = \frac{2}{16} = \frac{1}{8}$

Section 1.3 Exercises

Simplify.

1. $1+8-4\cdot 5$
2. $-2\div|-5|\cdot 10$
3. $1+2\cdot 3-4\div 2$
4. $4-6^2(5-17)$
5. $18\div(3^2-6)$
6. $\frac{4}{3}+2|-4-7|$
7. $0.6+3^2(1-(1-2))$
8. $4-7[3-2(5-6)+2^2]$
9. $2-18\left\{\left(\frac{1}{2}\right)^2\cdot 4^2+1\right\}$
10. $\left(-\frac{3}{5}\right)^2+\left(\frac{1}{5}\right)^3$
11. $\left(\frac{2}{3}-1^2\right)\left(4^2-\frac{2}{3}\right)$
12. $\frac{4-5\cdot 2}{18\div 9+7\cdot 3-2}$
13. $\frac{5-6(3-2^2)}{14-12\div 3}$

1.4: ORDER OF OPERATIONS—LIKE TERMS AND GROUPING SYMBOLS

In this section we will use like terms to simplify mathematical expressions. After that, we will look at how grouping symbols interact with these expressions by use of a major property from Section 1.1.

Like Terms

We begin with notation that will make the discussion of these and further topics a lot easier. An expression is a mathematical grouping of numbers and symbols with no equal sign. For example, the absolute value of x, $|x|$, is an expression.

Ex) $x^2 + 5x + 6$ is an expression.

Ex) $1 + 2 \div 3$ is an expression.

Now, in an expression there are terms. Terms are separated by addition. For example, in the expression $x^2 + 5x + 6$, the three terms are x^2, $5x$, and 6.

Even if we are subtracting, we separate the terms with addition.

Ex) Consider $1 + 2x - x^3$. Using the rule that subtraction is adding opposites, we think of this as $1 + 2x + (-x^3)$. So the terms of $1 + 2x - x^3$ are 1, $2x$, and $-x^3$.

YOU TRY IT: Write down the terms of $x^4 - x^3 + 1$.

We will be spending a lot of time in this text simplifying expressions. So it is crucial to know *when* expressions can be simplified. It stands to reason that if you have $x^2 + x^2$, you are adding one x^2 to another x^2, so you have two of them. Indeed, like terms have the same variable (or variables) raised to the same power. If a constant is multiplied by these variables, we call the resulting number the coefficient of the term.

Ex) In the expression $x^2 + x^3 - 2x^2 + 4x^3$, one set of like terms is x^2 and $-2x^2$. Also note that x^3 and $4x^3$ are like terms. The coefficient of x^2 is 1, the coefficient of $-2x^2$ is -2, and the coefficient of $4x^3$ is 4.

YOU TRY IT: Write down the terms of $2x^2 + 3xy - y^2 + 6y$ and their coefficients.

When we are adding like terms, we add the coefficients of the terms. This way, if we are adding $x^2 + x^2$, we are adding one x^2 to another x^2, which makes $2x^2$. This is all thanks to the distributive property, which we covered in Section 1.1. Remember that the distributive property tells us that $a(b+c) = ab + ac$. Well we are using this property from the right side to the left side. Think of it as

$$ba + ca = (b + c)a.$$

Then, using words, think of this as "b copies of a plus c copies of a." Then we have, in total, "b + c copies of a."

Example: Simplify $x^2 + x^3 - 2x^2 + 4x^3$.

Solution: We have

$$\begin{aligned} x^2 + x^3 - 2x^2 + 4x^3 &= x^2 - 2x^2 + x^3 + 4x^3 \\ &= (1-2)x^2 + (1+4)x^3 \\ &= (-1)x^2 + (5)x^3 \\ &= -x^2 + 5x^3. \end{aligned}$$

Here we are using the convention that -1 times a variable is the opposite of that variable. That is, $(-1)x^2$ is the same thing as $-x^2$.

Ex) To simplify $4y + x - 7y$, note that the y's are like terms, but the x must be left alone. So we can add the coefficients 4 and -7 to get -3. Thus, $4y + x - 7y = -3y + x$. Since addition is commutative, the answer $-3y + x$ is the same as the answer $x - 3y$.

When we begin to simplify complicated expressions, keep track by making marks under the like terms as you simplify them.

Example: Simplify $2y - 2x + 6x^2 - 3y + 3x - 8x^2$.

Solution: To keep things clear, let's underline the like terms that we are adding in each step. We have

$$\begin{aligned} \underline{2y} - 2x + 6x^2 - \underline{3y} + 3x - 3x^2 &= -y - \underline{2x} + 6x^2 + \underline{3x} - 8x^2 \\ &= -y + x + \underline{6x^2} - \underline{8x^2} \\ &= -y + x - 2x^2. \end{aligned}$$

Example: Simplify $3x^2 + 6xy - 5y^2 - 15xy + 10y^2 - 8x^2$ by combining like terms.

Solution: Underlining like terms as we go, we have

$$\begin{aligned} \underline{3x^2} + 6xy - 5y^2 - 15xy + 10y^2 - \underline{8x^2} &= -5x^2 + \underline{6xy} - 5y^2 - \underline{15xy} + 10y^2 \\ &= -5x^2 - 9xy - \underline{5y^2} + \underline{10y^2} \\ &= -5x^2 - 9xy + 5y^2. \end{aligned}$$

Example: Simplify $3x^2 + 7x^2y - 15y^2 + 13xy^2$ by combining like terms.

Solution: Take a second to look at the variables. We have an x^2 term, an x^2y term, a y^2 term, and an xy^2 term (with various coefficients). But there are no like terms. This expression is as simplified as it can be.

Example: Simplify $3x^2 + 7x^2y + 6xy^2 - 8xy^2 - 15x^2 + 3xy^2$.

Solution: Let's underline the like terms as we simplify. We have

$$\begin{aligned} 3x^2 + 7x^2y + 6xy^2 - 8xy^2 - 15x^2 + 3xy^2 &= \underline{3x^2} + 7x^2y + 6xy^2 - 8xy^2 - \underline{15x^2} + 3xy^2 \\ &= -12x^2 + 7x^2y + \underline{6xy^2} - \underline{8xy^2} + \underline{3xy^2} \\ &= -12x^2 + 7x^2y + xy^2. \end{aligned}$$

Example: Simplify $\frac{1}{2}x^3 - 0.3x^2 + \frac{2}{5}x^3 + 1.6x^2$.

Solution: We add the coefficients of x^3 together. So we make a side note that

$$\frac{1}{2} + \frac{2}{5} = \frac{5}{10} + \frac{4}{10} = \frac{5+4}{10} = \frac{9}{10}.$$

So we have

$$\begin{aligned} \frac{1}{2}x^3 - 0.3x^2 + \frac{2}{5}x^3 + 1.6x^2 &= \underline{\frac{1}{2}x^3} - 0.3x^2 + \underline{\frac{2}{5}x^3} + 1.6x^2 \\ &= \frac{9}{10}x^3 - \underline{0.3x^2} + \underline{1.6x^2} \\ &= \frac{9}{10}x^3 + 1.3x^2. \end{aligned}$$

Grouping Symbols

We can extend the distributive property to more than two numbers. For example, if we need to multiply 2(3 + 5 + 8), we can add $2 \cdot 3 + 2 \cdot 5 + 2 \cdot 8$. In fact, the distributive property will work for any number of values:

$$a(b+c+d+e+\cdots+z) = ab+ac+ad+ae+\cdots+az.$$

For emphasis, we usually draw arrows from the number we are distributing to each term in the parentheses.

Example: Simplify $-4(3+5+7)$.

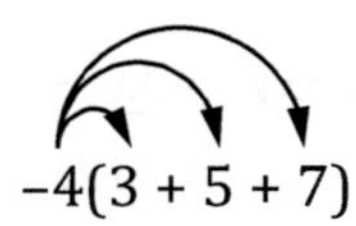

Fig. 1-5

Solution: This is $-4 \cdot 3 + (-4) \cdot 5 + (-4) \cdot 7 = -12 - 20 - 28 = -32 - 28 = -60$. See Figure 1-5.

Example: Simplify $-(8 - 2)$.

Solution: If we distribute a negative sign, we imagine distributing a -1. So we have

$$\begin{aligned} -(8-2) &= -1(8-2) \\ -1\cdot 8 - 1\cdot(-2) \\ &= -8 - 1(-2) \\ &= -8 + 2 \\ &= -6. \end{aligned}$$

This is an interesting take on simplifying. Note that when we discussed order of operations in the previous section, we demanded simplifying inside the parentheses first. This is one of the nice things about mathematics: There is often more than one way to find the solution to a problem.

Of course, we need to be able to distribute when we have variable expressions. For example, if we wanted to double $x + y$, the expression would be $2(x + y)$. By the distributive property, this is $2x + 2y$. So note that when you distribute a number to an expression, you are multiplying that number by the *coefficients* of the terms of that expression.

Example: Simplify $-3(x + 2y)$.

Solution: Distributing -3, we have $-3x + (-3)(2y) = -3x - 6y$.

If the term has a coefficient of -1, we again use the convention that there is an imaginary -1 in front of the term.

Example: Simplify $2(b - ab - 15b^2)$.

Solution: Distributing the 2 gives $2b + 2(-1ab) + 2(-15b^2) = 2b - 2ab - 30b^2$.

Example: Simplify $\frac{1}{2}\left(4x^2 - \frac{1}{2}x + \frac{4}{9}\right)$.

Solution: We have

$$\begin{aligned}\frac{1}{2}\left(4x^2-\frac{1}{2}x+\frac{4}{9}\right)&=\frac{1}{2}(4x^2)+\frac{1}{2}\left(-\frac{1}{2}x\right)+\frac{1}{2}\cdot\left(\frac{4}{9}\right)\\&=\frac{1}{2}\cdot\frac{4}{1}x^2-\frac{1}{2}\cdot\frac{1}{2}x+\frac{1}{2}\cdot\frac{4}{9}\\&=2x^2-\frac{1}{4}x+\frac{2}{9}.\end{aligned}$$

Example: Simplify $0.2(4x-3y^2)$.

Solution: Distributing the 0.2, we have $0.2(4x)+0.2(-3y^2)=0.8x-0.6y^2$.

YOU TRY IT: Simplify $14\left(2x-3y+\frac{1}{7}z\right)$.

Example: Simplify $-4(3-2(x-y))$.

Solution: We are combining all our knowledge of the order of operations here. Start from the inside, and distribute -2:

$$\begin{aligned}-4(3-2(x-y))&=-4(3-2\cdot x-2(-y))\\&=-4(3-2x+2y)\\&=-4\cdot 3-4(-2x)-4(2y)\\&=-12+8x-8y.\end{aligned}$$

YOU TRY **Solution:** Terms are $2x^2$, $3xy$, $-y^2$, and $6y$; coefficients are 2, 3, −1, 6.

Section 1.4 Exercises

1. a. What is an expression?
   b. What is a term?
   c. What is a coefficient?
2. a. How many terms are in the expression $3x^3+4x^2-5x+6$?
   b. List the terms and their coefficients.
3. a. How many terms are in the expression $0.3x-5y+2z$?
   b. List the terms and their coefficients.

4. Simplify $2x+3y-5x-10y$.
5. Simplify $\frac{1}{2}x+\frac{1}{3}y-\frac{1}{5}x-\frac{1}{10}y$.
6. Simplify $0.3x+1.2y-2x-0.5y$.
7. Simplify $2x-x^2+3x+2x^2-5$.
8. Simplify $x^2+2x+1+y^2+2y+7-3x^2-2y$.
9. Simplify $\frac{1}{2}x^2+4y^2-\frac{3}{2}x^2-y^2$.
10. Simplify $4x^2y-3xy^2+15y^2+10xy^2-8y^2+2x^2y$.
11. Simplify $4ab+2b-8a+10ab-18a+16b$.
12. Simplify $\frac{-49x-3y(7-4)}{y+3(2-y)}$.
13. Simplify $\frac{1}{8}(5-3)$.
14. Simplify $7(4x-3y)$.
15. Simplify $-2\left(x+\frac{1}{2}y\right)$.
16. Simplify $-2(3-5x(a+2))$.
17. Simplify $x(5-3^2)$.
18. Simplify $2-[1-3(y-x)]$.
19. Simplify $\frac{1}{2}\left[4-\frac{1}{3}(x-2)\right]$.

1.5: Rules for Exponents

In this section we will discuss the rules for exponents, which are used in every math class—and are the cause of many common errors. We will discuss expressions that have exponents that are integers.

Integer Exponents

For the expression a^m, we call the number a the base, and we call m the exponent. As an example, in the expression 2^3, the base is 2 and the exponent is 3.

In Section 1.3, we talked about natural number exponents. So we know that if n is positive, then

$$x^n = x \cdot x \cdot \dots \cdot x \ (n \text{ times}).$$

But what happens if we raise a number to the "zero power"? We have to choose a convention here, and because raising a number to a positive power deals with multiplication, it makes sense that when we raise a number to the zero power we get the multiplicative identity.

Any nonzero number raised to the zero power is 1.

Ex) We write $2^0 = 1$, and $(-2)^0 = 1$. Note carefully, however, that -2^0 is the opposite of 2^0, which is -1.

Ex) Similarly with variables, we have $x^0 = 1$, as long as $x \neq 0$, and we have $(a + b)^0 = 1$, as long as $a + b \neq 0$.

Now, in keeping with the ideas of multiplication and the multiplicative identity, we will define what negative exponents are:

Any nonzero number raised to the -1 power is its reciprocal.

Ex) We have $2^{-1} = \frac{1}{2}$.

Ex) Note that if we raised 0 to the -1 power we would have $0^{-1} = \frac{1}{0}$, and division by zero is not allowed. Both 0^{-1} and $\frac{1}{0}$ are undefined.

Example: Simplify $3(4+2^{-1})$.

Solution: Simplify the inside first, to get $3(4+2^{-1}) = 3\left(4+\frac{1}{2}\right) = 3\left(\frac{8}{2}+\frac{1}{2}\right) = 3\left(\frac{9}{2}\right) = \frac{3}{1}\cdot\frac{9}{2} = \frac{27}{2}$.

Note that we could also distribute the 3 like we did in Section 1.4.2.

If we understand a negative exponent to mean "reciprocal," we need to take one more step to make our definition of exponents for all integers. That is,

$$x^{-n} = \frac{1}{x^n}, x \neq 0, n \in \mathbb{Z}.$$

Example: Simplify $2^{-2} - 3^{-2}$.

Solution: $\frac{1}{2^2} - \frac{1}{3^2} = \frac{1}{4} - \frac{1}{9} = \frac{1}{4} \cdot \frac{9}{9} - \frac{1}{9} \cdot \frac{4}{4} = \frac{9-4}{36} = \frac{5}{36}$

What happens when a fraction is raised to a negative number? The rule states that a number raised to the -1 power is its reciprocal, and the reciprocal of a fraction switches the numerator and the denominator. Thus, for any fraction $\frac{a}{b}$, we have

$$\left(\frac{a}{b}\right)^{-1} = \frac{b}{a}.$$

Next, if we take a fraction to an integer less than -1, we first take care of the reciprocal and then simplify as normal.

Example: Simplify $\left(\frac{2}{3}\right)^{-2} - 3^{-2}$.

Solution: We need to make the exponents positive so we can start simplifying, so we take reciprocals first:

$$\begin{aligned}\left(\frac{2}{3}\right)^{-2} - 3^{-2} &= \left(\frac{3}{2}\right)^{2} - \frac{1}{3^2} \\ &= \frac{3^2}{2^2} - \frac{1}{9} \\ &= \frac{9}{4} - \frac{1}{9} \\ &= \frac{9}{4} \cdot \frac{9}{9} - \frac{1}{9} \cdot \frac{4}{4} \\ &= \frac{81}{36} - \frac{4}{36} \\ &= \frac{77}{36}.\end{aligned}$$

YOU TRY IT: Simplify $\left(\frac{2}{5}\right)^{-2}+\left(\frac{2}{3}\right)^{-1}$.

Exponent Rules

Working with the above definitions, we can handle the rest of the rules for exponents.

When multiplying two expressions that have the same base, we add the exponents:

$$a^m a^n = a^{m+n}.$$

If we multiply m copies of a, and then we multiply that number by n copies of a, it stands to reason that there are a total of $m + n$ copies of a being multiplied together.

Ex) $2^2 \cdot 2^3 = 2^{2+3} = 2^5 = 32$. It always helps to imagine it when you can. We have two copies of 2 multiplied by 3 copies of 2, so there are 5 copies:

$$2^2 \cdot 2^3 = (2 \cdot 2) \cdot (2 \cdot 2 \cdot 2) = 2 \cdot 2 \cdot 2 \cdot 2 \cdot 2 = 2^5 = 32.$$

Ex) $x^4 x^6 = x^{4+6} = x^{10}$

Ex) $y^2 y^{-7} = y^{2+(-7)} = y^{-5} = \frac{1}{y^5}$

Example: Simplify $x \cdot x^3 \cdot x^5$.

Solution: If no exponent is written, we assume an exponent of 1. So we have

$$x^1 x^3 x^5 = x^{1+3+5} = x^9.$$

Ex) If $x \neq 0$, let's simplify $x \cdot x^{-1}$. Since we have the same base, we add the exponents to get $x^{1+(-1)} = x^0 = 1$. This concurs with our definition of raising a number to the -1 power, because we have $x \cdot x^{-1} = x \cdot \frac{1}{x} = \frac{x}{x} = 1$.

When dividing two expressions that have the same base, we subtract the exponents:

$$\frac{a^m}{a^n} = a^{m-n}.$$

Ex) $\frac{2^2}{2^3} = 2^{2-3} = 2^{-1} = \frac{1}{2}$. This is the same thing as canceling when you have

$$\frac{2^2}{2^3} = \frac{2 \cdot 2}{2 \cdot 2 \cdot 2} = \frac{1}{2}.$$

Example: Simplify $\frac{x^{64}}{x^{57}}$.

Solution: Since we are dividing two numbers with the same base, we subtract exponents, so

$$\frac{x^{64}}{x^{57}} = x^{64-57} = x^7.$$

Ex) Simplifying $\frac{x^2}{x^{-2}}$, we have to be careful of the negative exponent in the denominator. So using our division rule results in $\frac{x^2}{x^{-2}} = x^{2-(-2)} = x^{2+2} = x^4$.

We follow the convention to *always leave our answers with positive exponents*. It is usually helpful (that is, easier) to think of positive exponents rather than negative ones.

Working with a negative exponent means we need to start by taking the reciprocal. Taking care of the negative exponent corresponds to moving something from numerator to denominator, or from denominator to numerator. For example, when writing $\frac{1}{x^{-1}}$, we can remove the negative exponent if we move x from the denominator to the numerator. So $\frac{1}{x^{-1}} = x^1 = x$.

Ex) We can simplify $\frac{x^{-6}}{y^{-5}}$ by moving x^6 downstairs and y^5 upstairs. So $\frac{x^{-6}}{y^{-5}} = \frac{y^5}{x^6}$.

Example: Simplify $x(x^2 + xy^2)$.

Solution: We cannot simplify inside of the parentheses, so we must distribute the x, as follows:

$$\begin{aligned} x(x^2 + xy^2) &= x(x^2) + x(xy^2) \\ &= x \cdot x^2 + x \cdot x \cdot y^2 \\ &= x^{1+2} + x^{1+1}y^2 \\ &= x^3 + x^2y^2. \end{aligned}$$

Example: Simplify

$$\frac{x^{-5}}{x^{-7}} \cdot \frac{x^3}{x^4}.$$

Solution: There are a number of ways to solve this. But let's multiply fractions, and use our exponent rules. We have

$$\begin{aligned}\frac{x^{-5}}{x^{-7}} \cdot \frac{x^3}{x^4} &= \frac{x^{-5} \cdot x^3}{x^{-7} \cdot x^4} \\ &= \frac{x^{-5+3}}{x^{-7+4}} \\ &= \frac{x^{-2}}{x^{-3}} \\ &= x^{-2-(-3)} \\ &= x^{-2+3} \\ &= x.\end{aligned}$$

YOU TRY IT: Try simplifying the previous example by taking care of the negative exponents first. That is, move things "upstairs" and "downstairs," and then multiply the fractions.

Example: Simplify $(2x^2)(-3xy^3)(4x^2y)$.

Solution: Note that these terms in parentheses are not "like" terms, so we cannot add them together. But we certainly can multiply them together. We move the coefficients, x's, and y's around so they are grouped together:

$$\begin{aligned}(2x^2)(-3xy^3)(4x^2y) &= 2 \cdot (-3) \cdot 4 \cdot x^2 \cdot x \cdot x^2 \cdot y^3 \cdot y \\ &= -24x^{2+1+2}y^{3+1} \\ &= -24x^5y^4.\end{aligned}$$

Example: Simplify $-2x^{-1}(4x^3+3x^2-10x-5)$.

Solution: When distributing here, be very careful. Be liberal with parentheses! A negative coefficient and a negative exponent mean different things (they are opposite and reciprocal, respectively). We have

$$\begin{aligned}-2x^{-1}(4x^3+3x^2-10x-5) &= -2x^{-1}(4x^3)-2x^{-1}(3x^2)-2x^{-1}(-10x)-2x^{-1}(-5)\\ &= -2\cdot 4x^{-1+3}-2\cdot 3x^{-1+2}-2(-10)x^{-1+1}-2(-5)x^{-1}\\ &= -8x^2-6x^1+20x^0+10x^{-1}\\ &= -8x^2-6x+20+\frac{10}{x}.\end{aligned}$$

In the transition to the last step here, we are being careful with notation: We are not leaving an exponent of 1, something to the zero power is 1, and we make sure to have a positive exponent.

The last rule answers this question: What happens when we raise something to a power that already has an exponent? For example, what is $(x^2)^5$? Well, we are raising x^2 to the fifth power, so this means that we are multiplying 5 copies of x^2 together:

$$x^2\cdot x^2\cdot x^2\cdot x^2\cdot x^2 = x^{2+2+2+2+2} = x^{10}.$$

The rule is to multiply those exponents of 2 and 5. Thus,

$$(a^m)^n = a^{mn},\ a\in\mathbb{R},\ \ m,n\in\mathbb{Z}.$$

Ex) $(2^2)^3$ is the cube of 2^2, so we can write this as $2^{2\cdot 3}=2^6=64$.

Ex) The same rule applies to a product being raised to a power. If we have $(x^2y^2)^3$ this is 3 copies of x^2y. So we multiply 3 by all of the exponents: $(x^2y)^3=(x^2)^3(y)^3=x^{2\cdot 3}y^{1\cdot 3}=x^6y^3$.

Example: Simplify $x^3(2x^{-2}y^5)^4$.

Solution: Remembering the order of operations, we must apply the exponent of 4 first:

$$\begin{aligned} x^3(2x^{-2}y^5)^4 &= x^3 \cdot 2^4(x^{-2})^4(y^5)^4 \\ &= x^3 \cdot 2^4 \cdot x^{-2\cdot 4} \cdot y^{5\cdot 4} \\ &= x^3 \cdot 16 \cdot x^{-8} \cdot y^{20} \\ &= 16 \cdot x^3 \cdot x^{-8} \cdot y^{20} \\ &= 16x^{-5}y^{20}. \end{aligned}$$

Note that the multiplication symbol in the last step is suppressed, because multiplication is implied.

The following is a summary of the rules in this section.

In words	In mathematical symbols
Anything to the zero power is 1 (except 0).	$a^0 = 1, a \neq 0$
A negative exponent means reciprocal.	$a^{-n} = \frac{1}{a^n}, a \neq 0$
When *multiplying* two expressions with the same base, we *add* the exponents.	$a^m a^n = a^{m+n}$
When *dividing* two expressions with the same base, we *subtract* the exponents.	$\frac{a^m}{a^n} = a^{m-n}$
When raising an expression with an exponent to another exponent, we multiply the exponents.	$(a^m)^n = a^{mn}$

Section 1.5 Exercises

Simplify, assuming no division by zero.

1. a^0, where a is a positive number
2. $(a+b+c+d)^0$, where $a+b+c+d \neq 0$
3. $2^0 - 3^0$
4. $5^{-1} - 6^{-1}$

5. $x^0 + \left(\frac{1}{4}\right)^{-1}$

6. $5\left(w^0 - \left(\frac{1}{5}\right)^{-1}\right)$

7. $5^{-2} - 7^{-2}$

8. $\left(\frac{3}{7}\right)^{-2} + \left(\frac{9}{2}\right)^{-2}$

9. $4\left(\frac{3}{2}\right)^{-3} - \left(\frac{2}{x}\right)^{-2}$

10. x^2x^{18}

11. $x^2(x + 2y)$

12. $3x(4 - 7y - x^{-1})$

13. $2xy(-3x^2y^4)$

14. $x^2[x - y(3 - x)]$

15. $x^3[4 - x^2(5 - x) + 7]$

16. $\frac{1}{2}a^2(ab - a^3)$

17. $x^{-2}\left(5 - \frac{x^3}{x^2}\right)$

18. $\frac{x^6}{x^{63}}$

19. $\frac{x^{-5}y^{-2}}{y^{-8}x^{-3}}$

20. $\frac{x^2(x^{-3}x)}{y^2y^{-2}}$

21. $(x^2)^3$

22. $(x^2)^{-3}$

23. $(x^{-2})^{-3}$

24. $\frac{(x^{-2})^{-5}}{(x^2)^3}$

25. $\frac{(2x^2yz^4)^3}{(4x^{-3}y^2z^5)^2}$

26. $\frac{(9w^{-3}z^{-4})^{-2}}{(3w^{-2}z^{-5})^{-3}}$

1.6: Using Formulas

In this section we will be evaluating formulas where the variables are taken to be specific numerical values. First up, we need to discuss evaluating expressions.

Evaluating Expressions

An expression allows variables to be anything within reason. So when evaluating expressions, the most important thing to remember is to *use parentheses when plugging in the value.*

Example: Evaluate x^2 when $x = 2$.

Solution: We have $(2)^2 = 4$.

It seems easy here, but the next example shows why we need the parentheses.

Example: Evaluate $-x^2$ in these instances: (a) when $x = 2$ and (b) when $x = -2$.

Solution:

(a) Remember that $-x^2$ is the opposite of the square of x. We need parentheses around the variables we are substituting. So $-(2)^2 = -4$.

(b) When $x = -2$,

$$-(-2)^2 = -(4) = -4.$$

Example: Evaluate $-x^2 - x + \frac{2}{7}$ when $x = -1$.

Solution: We have

$$\begin{aligned} -(-1)^2 - (-1) + \frac{2}{7} &= -(1) + 1 + \frac{2}{7} \\ &= 0 + \frac{2}{7} \\ &= \frac{2}{7}. \end{aligned}$$

Example: Evaluate $\frac{2}{x^3} - x^2$ when $x = -4$.

Solution: We substitute in using parentheses, and obtain

$$\begin{aligned} \frac{2}{(-4)^3} - (-4)^2 &= \frac{2}{-64} - 16 \\ &= -\frac{1}{32} - 16 \\ &= -\frac{1}{32} - \frac{512}{32} \\ &= -\frac{513}{32}. \end{aligned}$$

Note how we moved the negative sign out front when we went from the first line to the second line in our solution.

Formulas

Now that we can handle expressions, formulas are just an extension of what we already know. A formula has an equal sign in it. Some examples are $A = s^2$ or $d = rt$. It's important to note the difference between an expression and a formula because we have different rules for expressions and equations. Expressions do not have an equal sign, whereas formulas and equations do. (We will discuss this difference more in future sections.)

The formula $d = rt$ is the formula that relates "distance d traveled at a constant rate r over a time period t."

Ex) Suppose we travel for four hours at a constant speed of 50 miles per hour. Then the time $t = 4$ hours, and the rate $r = 50$ mph. Using the formula, the distance we've traveled is

$$d = rt = (4\,\text{hr.})\left(50\,\frac{\text{mi.}}{\text{hr.}}\right) = 200\,\text{mi.}$$

Notice the canceling of the units of "hours" in the previous example, so that only "miles" is left behind. We can use units to help us make sense about what a variable's value should be.

The formula $A = s^2$ gives the area A of a square with side length s.

Example: Find the area of a square with side length 3.5 feet.

Solution: Since the side length is $s = 3.5$ feet, we have

$$\begin{aligned} A &= (3.5\,\text{ft.})^2 \\ &= 3.5^2\,\text{ft.}^2 \\ &= 12.25\,\text{ft.}^2 \end{aligned}$$

It should make sense that an *area* is in square feet, where as a side *length* is in feet.

An exponent is sometimes called a **superscript**. If we write a number below and to the right of a variable, it is called a **subscript**. This practice is helpful in naming similar variables. For example, a_1 is read as "a one" or "a sub one."

The total resistance R_T of two resistors R_1 and R_2 connected in parallel is given by the formula

$$R_T = \left(\frac{1}{R_1} + \frac{1}{R_2}\right)^{-1}.$$

Example: Two resistors R_1 and R_2 are connected in parallel. If resistor R_1 has 3 ohms of resistance, and resistor R_2 has 4 ohms of resistance, what is the total resistance?

Solution: We have $R_1 = 3$, and $R_2 = 4$, so the total resistance is

$$\begin{aligned} R_T &= \left(\frac{1}{3}+\frac{1}{4}\right)^{-1} \\ &= \left(\frac{4}{12}+\frac{3}{12}\right)^{-1} \\ &= \left(\frac{7}{12}\right)^{-1} \\ &= \frac{12}{7}. \end{aligned}$$

The total resistance is $\frac{12}{7}$ ohms.

The <u>average</u> or <u>mean</u> $\overline{x}$ (read "x bar") of n numbers $x_1, x_2, \ldots, x_n$ is given by

$$\overline{x} = \frac{x_1 + x_2 + \cdots + x_n}{n}.$$

This means we take the sum of the numbers, and then divide that by how many numbers we have.

Example: Find the average of 3, −6, 15, and 7.

Solution: There are four numbers here, so $n = 4$. The average is

$$\begin{aligned} \overline{x} &= \frac{3+(-6)+15+7}{4} \\ &= \frac{19}{4}, \end{aligned}$$

or 4.75.

Section 1.6 Exercises

1. Evaluate $\frac{1}{x} + 4x$ when $x = 2$.
2. Evaluate $-x^2 - 18x - 15$ when $x = -1$.
3. Evaluate $\frac{x^2 y^3}{x^2 + y^3}$ when $x = 1$ and $y = 4$.
4. Evaluate $\frac{x^2 y^3}{x^2 + y^3}$ when $x = -1$ and $y = -2$.
5. Evaluate $y(3 - x^2)$ when $x = -2$ and $y = -3$.
6. An equilateral triangle is a triangle whose side lengths are all the same. The area of an equilateral triangle with side length s is $A = \frac{\sqrt{3}}{4} s^2$.
   a. Find the area of an equilateral triangle with side length 2 cm.
   b. Find the area of an equilateral triangle with side length $\frac{1}{2}$ cm.
7. Use the formula for total resistance to find the total resistance when $R_1 = 5$ ohms and $R_2 = 8$ ohms.
8. Use the formula for total resistance to find the total resistance when $R_1 = \frac{1}{4}$ and $R_2 = \frac{1}{2}$ ohms.
9. Use the formula for the mean to find the average of $1, -15, 32, 20$, and -10.
10. Use the formula for the mean to find the average of $-3, -6, 19$, and 10.
11. Looking ahead: The *discriminant* is used to describe the solutions of quadratic equations. The discriminant of the quadratic $ax^2 + bx + c$ is $b^2 - 4ac$.
    a. Calculate the discriminant of $2x^2 - x + 1$.
    b. Calculate the discriminant of $\frac{1}{2}x^2 + \frac{1}{2}x - 7$.
12. Looking ahead: The kinetic energy of an object in motion is $K = \frac{1}{2}mv^2$, where m is the object's mass and v is the object's velocity. Find the kinetic energy of a 5-kilogram object moving at $15\frac{\text{m}}{\text{sec}}$. The units for kinetic energy are **joules**, which are kilograms times meters² divided by seconds², or $\text{J} = \text{kg} \cdot \frac{\text{m}^2}{\text{sec}^2}$.

Chapter 1 Review

1-A

1. Of real number, rational number, irrational number, integer, and natural number, which of these can be applied to $\sqrt{2}$?
2. Of real number, rational number, irrational number, integer, and natural number, which of these can be applied to $1.\overline{7}$?
3. Simplify $14 + 1 \cdot 2$.
4. Simplify $2^2 + 3^2$.
5. Simplify $x^2 + 3x + 5x^2 + 6x$.
6. Distribute to simplify $2(x - 3)$.
7. Simplify x^2x^3.
8. Evaluate $x^2 + x + 1$ when $x = -1$.

1-B

1. Of real number, rational number, irrational number, integer, and natural number, which of these can be applied to π?
2. Of real number, rational number, irrational number, integer, and natural number, which of these can be applied to $\pi - \pi$?
3. Simplify $14 - 1 \cdot 2 + 4 \div 3$.
4. Simplify $\left(-\frac{2}{3}\right)^2 + \left(-\frac{3}{4}\right)^2$.
5. Simplify $x^2y^2 + 4xy^2 + 8x^2y - 10xy^2 - 7x^2y + 3x^2y^2$.
6. Simplify $2[3 - (5 - x)]$.
7. Simplify $\frac{x^4}{y^7} \cdot \frac{y^3}{x^{-6}}$.
8. Evaluate $x^2y + xy^2 + x + y$ when $x = -1$ and $y = -1$.

CHAPTER

2

Linear Equations

In this chapter we will discuss the methods for solving linear equations for one variable. The main rule for solving equations is that whatever you do to one side of the equation, you have to do to the other side.

Another thing to think about is what exactly we are doing with an equation. If we're asked to solve $x + 3 = 4$, there is an implied question here. You are asking "For what x is it true that $x+3=4$?" And that is sometimes not a trivial question. For example, asking "For what x is it true that $x + 3 = 4$?" is very different than asking "For what x is it true that $x + 3 > 4$?" The equation $x + 3 = 4$ has one solution, while the inequality (which we will cover in Chapter 4) has *infinitely many* solutions. For the most part, however, we can think of all the equations in this text as having solutions in the real numbers. So usually, we think of the question as being "For what value(s) of $x \in \mathbb{R}$ is it true that $x + 3 = 4$?"

In this chapter we will solve equations for one variable, whether it is the only variable in the equation (covered in Section 2.1) or not (Section 2.2). We will also discuss absolute value equations (Section 2.3) and some useful applications (Section 2.4).

2.1: Equations with One Variable

In this section we will cover the rules for solving equations. We will be undoing the operations performed on the variable in order to isolate it.

One-Sided Equations

The operations of addition and subtraction are inverses. To solve $x + 3 = 4$, we see that 3 is being added to x. In order to isolate the variable x, we subtract 3 from both sides. We have

$$\begin{aligned} x+3&=4 \\ x+3-3&=4-3 \\ x&=1. \end{aligned}$$

The same strategy is used if we subtract a number from the variable.

Ex) To solve $x - 5 = -1$, we add 5 to both sides, which will undo the subtraction being done:

$$x-5+5=-1+5$$
$$x=4.$$

Recall that if there is a coefficient of the variable, this number is multiplied by the variable. For example, in solving $3y = 4$, we see that y has a coefficient of 3. To undo the multiplication, we divide by the coefficient on both sides.

Ex) In solving $3y = 4$, we have

$$3y=4$$
$$\frac{3y}{3}=\frac{4}{3}$$
$$y=\frac{4}{3}.$$

Notice that the 3's cancel on the left. Remember when we cancel numbers, there is a 1 leftover. So on the left we have $\frac{3y}{3}=\frac{1y}{1}=y.$

Example: Solve $\frac{z}{7}=\frac{1}{2}$.

Solution: Since z is being divided by 7, we can undo this by multiplying both sides by 7:

$$7\cdot\frac{z}{7}=7\cdot\frac{1}{2}.$$

Again note that the 7's will cancel on the left-hand side, leaving us with z, which is what we want. On the right we have $\frac{7}{1}\cdot\frac{1}{2}=\frac{7}{2}$. Thus our solution is $z=\frac{7}{2}$.

YOU TRY IT: (1) Solve $4w=\frac{9}{4}$.

The more difficult problems have coefficients that are fractions. Recall that dividing by a number is the same as multiplying by the reciprocal.

Ex) If we are solving $\frac{2}{3}x = 6$, we divide both sides by $\frac{2}{3}$, which is the same as multiplying both sides by $\frac{3}{2}$:

$$\frac{3}{2}\left(\frac{2}{3}x\right) = \frac{3}{2}(6)$$
$$x = \frac{3}{2} \cdot \frac{6}{1} = 9.$$

YOU TRY IT: (2) Solve $\frac{9}{4}q = 4$.

Two-Sided Equations

Most of the time when we are solving equations, we need to isolate the variable somehow. (There are exceptions to this rule, and we will cover that when we talk about quadratic equations in Chapter 8.) When the variable is on both sides of the equation, we need to get it to one side.

Example: Solve $3x - 2 = 4x + 1$.

Solution: We have a choice to subtract $3x$ from both sides, or to subtract $4x$ from both sides. Since dealing with positive coefficients is easier, let's subtract $3x$ from both sides:

$$\begin{aligned} 3x - 2 &= 4x + 1 \\ -3x \quad &\quad -3x \\ -2 &= x + 1 \\ -1 \quad &\quad -1 \\ -3 &= x. \end{aligned}$$

Let's check that this solution works. When we plug in -3 for x on both sides,

$$\begin{aligned} 3(-3) - 2 &= 4(-3) + 1 \\ -9 - 2 &= -12 + 1 \\ -11 &= -11. \end{aligned}$$

Example: Solve $2(x+1)+3=-4(5-x)-2$.

Solution: Sometimes we have to simplify both sides before getting x to one side. In the following series, let's distribute to get rid of the parentheses:

$$\begin{aligned}
2(x+1)+3 &= -4(5-x)-2 \\
2x+2+3 &= -4(5)-4(-x)-2 \\
2x+5 &= -20+4x-2 \\
2x+5 &= 4x-22 \\
-2x \quad & \quad -2x \\
5 &= 2x-22 \\
+22 \quad & \quad +22 \\
27 &= 2x \\
\frac{27}{2} &= \frac{2x}{2} \\
\frac{27}{2} &= x.
\end{aligned}$$

When we were three steps toward our solution above, we subtracted $2x$ from both sides (where the text is shown in blue). It's important to note that we could have subtracted $4x$ from both sides as well. Let's see that it works:

$$\begin{aligned}
2x+5 &= 4x-22 \\
-4x \quad & \quad -4x \\
-2x+5 &= -22 \\
-5 \quad & \quad -5 \\
-2x &= -27 \\
\frac{-2x}{-2} &= \frac{-27}{-2} \\
x &= \frac{27}{2}.
\end{aligned}$$

This is the same solution as above.

YOU TRY IT: (3) Solve $-(v+1)-2=2(v+3)$.

Example: Solve $\frac{1}{2}(x-6)=\frac{1}{4}(3-x)$.

Solution: Distributing the fractions on both sides, we get:

$$\frac{1}{2}x+\frac{1}{2}(-6)=\frac{1}{4}(3)+\frac{1}{4}(-x)$$
$$\frac{1}{2}x-3=\frac{3}{4}-\frac{1}{4}x.$$

Now let's add $\frac{1}{4}x$ to both sides:

$$+\frac{1}{4}x \qquad +\frac{1}{4}x$$
$$\left(\frac{1}{4}+\frac{1}{2}\right)x-3=\frac{3}{4}$$
$$\frac{3}{4}x-3=\frac{3}{4}$$
$$+3 \quad +3$$
$$\frac{3}{4}x=\frac{3}{4}+\frac{12}{4}$$
$$\frac{3}{4}x=\frac{15}{4}$$
$$\frac{4}{3}\cdot\frac{3}{4}x=\frac{15}{4}\cdot\frac{4}{3}$$
$$x=\frac{15}{3}=5.$$

YOU TRY IT Solutions:

(1) $x=\frac{9}{16}$; (2) $q=\frac{16}{9}$; (3) $v=-3$

Section 2.1 Exercises

Solve each equation.

1. $x+2=\frac{7}{2}$
2. $3x+1=\frac{4}{3}$
3. $0.4x+0.1=0.8$
4. $\frac{5}{3}w=6$
5. $\frac{2}{7}z=\frac{2}{7}$
6. $\frac{2}{9}p+1=\frac{13}{9}$
7. $\frac{1}{3}(w+4)=2$
8. $4(1-z)-5=-1$
9. $7x+1=3x+5$
10. $8x-6=4-2x$
11. $0.2x+4=0.5x$
12. $2(z+1)=4(3-z)$
13. $-6(h+4)=2(h-1)$
14. $\frac{1}{3}(q-6)=\frac{2}{3}(q-12)$
15. $\frac{1}{4}(8-z)=\frac{3}{4}(z+2)$
16. $0.1(10+100x)=0.3(10x-50)$
17. $0.2(x-1)=0.3(6-x)$
18. $3\left(4+\frac{2}{3}x\right)=8\left(\frac{1}{2}x+6\right)$
19. What happens when we solve the equation $3x-4=3x-5$? What value(s) of x make this equation true?
20. What happens when we solve the equation $2x-6=-(6-2x)$? What value(s) of x make this equation true?

2.2: Equations with More than One Variable

Section 1.6 dealt with formulas. Sometimes we need to solve for a different variable in a formula. For example, when using $d = rt$ (distance equals rate times time), perhaps we know how far we traveled and how long it took us. If the rate we traveled is constant, we would be looking to solve for r in the equation $d = rt$. In this section, we will take equations with more than one variable in them and solve for a particular variable.

Example: Solve $d = rt$ for r. If we traveled 100 miles in 2 hours at a constant rate, what is that rate?

Solution: Since we are solving for r, we isolate it. Since r is multiplied by t, divide by t on both sides:

$$d = rt$$
$$\frac{d}{t} = \frac{rt}{t}$$
$$\frac{d}{t} = r.$$

If we divide how far we went by how long it took us, we'll get the constant rate that we traveled. If we traveled $d = 100$ miles in $t = 2$ hours, then our constant rate is

$$r = \frac{100}{2}\frac{mi}{hr}$$
$$r = 50\frac{mi}{hr}.$$

Example: Suppose we traveled a distance of d miles while traveling at a constant rate of r miles per hour. How long did it take us?

Solution: In the formula $d = rt$, we know d and r, and we are looking for t. So now we want to solve the equation for t. Let's divide both sides by r:

$$d = rt$$
$$\frac{d}{r} = \frac{rt}{r}$$
$$\frac{d}{r} = t.$$

So it took us $\frac{d}{r}$ hours to travel a distance of *d* miles at a rate of *r* miles per hour.

Example: Use the previous formula to determine how long it would take to drive 300 miles at 60 miles per hour.

Solution: The formula solved for time is $t = \frac{d}{r}$. So if $d = 300$ miles, and $r = 60$ miles per hour, then the time is

$$t = \frac{300}{60} = 5.$$

So it would take us 5 hours if we drive at a rate of 60 miles per hour.

Example: The average of *a*, *b*, and *c* is

$$\frac{a+b+c}{3} = d.$$

Solve this formula for *b*.

Solution: When solving for *b*, treat the other variables like numbers. We are isolating *b*. We will start by multiplying both sides by 3.

$$\begin{aligned} 3\left(\frac{a+b+c}{3}\right) &= 3d \\ a+b+c &= 3d \\ -a \quad &\quad -a \\ b+c &= 3d-a \\ -c \quad &\quad -c \\ b &= 3d-a-c. \end{aligned}$$

Example: Over the semester, Jason took three exams in his class. He got a 60 on his first exam and an 80 on his last exam, but he can't remember what he got on the second exam. Use the previous formula to find his score on the second exam if his average score was 65.

Solution: Jason's first exam score is $a = 60$, his third exam score is $c = 80$, and his average is $d = 65$. So his second exam score, b, is given by

$$\begin{aligned} b &= 3d - a - c \\ b &= 3(65) - (60) - (80) \\ b &= 195 - 60 - 80 \\ b &= 55. \end{aligned}$$

Jason got a 55 on his second exam.

Example: Solve the equation

$$\frac{x + yz}{7} = \frac{w}{28}$$

for z.

Solution: To start isolating for z, we need to get rid of the 7 in the denominator. To do this, we'll multiply both sides by 7:

$$\begin{aligned} 7\left(\frac{x+yz}{7}\right) &= 7\left(\frac{w}{28}\right) \\ x + yz &= \frac{w}{4} \\ -x \quad & \quad -x \\ yz &= \frac{w}{4} - x \\ \frac{1}{y}(yz) &= \left(\frac{w}{4} - x\right)\cdot\frac{1}{y} \\ z &= \frac{w}{4}\left(\frac{1}{y}\right) - \frac{x}{1}\left(\frac{1}{y}\right) \\ z &= \frac{w}{4y} - \frac{x}{y} \end{aligned}$$

Section 2.2 Exercises

1. Solve $A = lw$ for l.
2. If the area of a square is 12 *feet*^{2} and the width is 3 feet, use the formula in the previous problem to find the length.
3. If we travel 500 miles in 10 hours at a constant speed, what is that speed?
4. How long will it take to make a 40-mile journey biking at 8 miles per hour?
5. The ideal gas law in Chemistry is $PV = nRT$, where P is the pressure of the gas, V is the volume of the gas, n is the amount of gas in moles, R is the universal gas constant, and T is the temperature of the gas. Solve this equation for the volume V.
6. Solve the ideal gas law $PV = nRT$ for the temperature T.
7. In Chemistry, Boyle's law states that the product of the volume of a gas V times the pressure of the gas P is a constant k. That is, $PV = k$. Solve this equation for V.
8. In Physics, the kinetic energy of a moving object is $K = \frac{1}{2}mv^2$. (See Section 1.6 Exercise 12.) Solve for the object's mass P.
9. An object has a kinetic energy of 100 J and is moving at 5 meters per second. Use your work in the previous exercise to find its mass m in kilograms.
10. In physics, Hooke's law states that the force F exerted on a spring is a constant k times the distance the spring is stretched beyond its natural state x. That is, $F = kx$. Solve this equation for k.
11. The unit for force is *newton*, abbreviated "N." If the force on a certain spring $F = 10$N is required to stretch the spring a distance of 20 meters, find the spring constant k for this spring using your work from Exercise 10 above.

2.3: Absolute Value Equations

Recall that in Section 1.2 we discussed absolute value. The *absolute value* of a number is its distance from the origin.

So if we ask ourselves "For what x is $|x| = 5$?," we are really asking for the numbers x whose distance from the origin is 5. There are two numbers that are 5 units from the origin: 5 and -5, so absolute value equations can have more than one solution.

Example: Solve $|3x| = 9$.

Solution: If the distance between the origin and $3x$ is 9, then $3x = 9$ or $-3x = 9$. See Figure 2-1. Solving these equations, we see that $x = 3$ or $x = -3$. Plugging each of these values back into the original equation shows that they are both solutions.

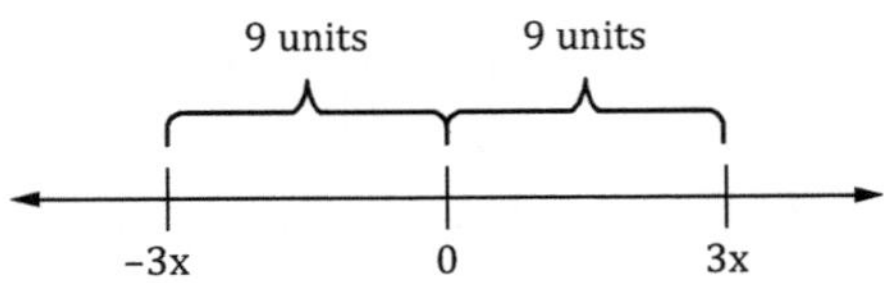

Fig. 2-1
Distance between the origin and 3x is 9.

In general, when we're solving absolute value equations, the strategy is first to isolate the *absolute value*, and then to write one equation without the bars and one where we take the opposite of one side of the equation. Usually we pick the side that is just a number. That is, if the equation is

$$|\text{what's inside}| = \text{nonnegative number},$$

then we will break this down into two equations:

$$\text{what's inside} = \text{nonnegative number}, \qquad \text{what's inside} = -(\text{nonnegative number}).$$

Remember that it is always a good idea to check that both solutions work.

Example: Solve $|2x + 1| = 4$.

Solution: We write two equations. One equation just lifts the absolute value bars. But remember to write another equation where we lift the absolute value bars *and* take the opposite of the other side of the equation. So we are solving $2x + 1 = 4$ and $2x + 1 = -4$:

$$\begin{array}{rcr} 2x+1=4 & & 2x+1=-4 \\ -1-1 & & -1 \quad -1 \\ 2x=3 & or & 2x=-5 \\ x=\dfrac{3}{2} & or & x=-\dfrac{5}{2} \end{array}$$

Let's check that $x = \frac{3}{2}$ works. We have

$$\left|2\left(\frac{3}{2}\right)+1\right| = |3+1| = |4|,$$

which is 4. Thus $x = \frac{3}{2}$ is a solution. Next, we have

$$\left|2\left(-\frac{5}{2}\right)+1\right| = |-5+1| = |-4|,$$

which is also 4. Thus $x = -\frac{5}{2}$ is also a solution.

Example: Solve $|2 - x| - 1 = 0$.

Solution: We need to isolate the absolute value before we can take another step. So we add 1 to both sides for

$$|2 - x| = 1$$

Now we are solving

$$\begin{gathered} 2-x=1 \quad or \quad 2-x=-1 \\ -2-2 \qquad -2 \quad -2 \\ -x=-1 \quad or \quad -x=-3 \\ -1\cdot(-x)=-1\cdot(-1)\ or -1(-x)=-1(-3) \\ x=1 \quad or \quad x=3. \end{gathered}$$

Let's check that these are both solutions. If $x = 1$, we have $|2 - 1| -1 = |1| - 1 = 1 - 1 = 0$, which solves the equation. If $x = 3$, we have $|2 - 3| -1 = |-1| - 1 = 1 - 1 = 0$. So both 1 and 3 are solutions.

Example: Solve $\frac{1}{2}-\frac{3}{4}|x+3|=-4$.

Solution: Our first step is to subtract $\frac{1}{2}$ from both sides in order to isolate the absolute value:

$$\begin{aligned}\frac{1}{2}-\frac{3}{4}|x+3|&=-4\\ -\frac{1}{2}\qquad\quad &\ \ -\frac{1}{2}\\ -\frac{3}{4}|x+3|&=-\frac{8}{2}-\frac{1}{2}\\ -\frac{3}{4}|x+3|&=-\frac{9}{2}.\end{aligned}$$

Now we multiply both sides by the reciprocal of $-\frac{3}{4}$:

$$\begin{aligned}-\frac{4}{3}\cdot-\frac{3}{4}|x+3|&=-\frac{4}{3}\left(-\frac{9}{2}\right)\\ |x+3|&=6\end{aligned}$$

Now we can write our two equations:

$$\begin{aligned}x+3=6 \quad &\text{or} \quad x+3=-6\\ -3\ -3 \quad &\qquad\ -3\ -3\\ x=3 \quad &\text{or} \quad x=-9.\end{aligned}$$

Check the work:

If $x=3$, we have $\frac{1}{2}-\frac{3}{4}|3+3|=\frac{1}{2}-\frac{3}{4}(6)=\frac{1}{2}-\frac{18}{4}=\frac{2}{4}-\frac{18}{4}=-\frac{16}{4}=-4$.

If $x=-9$, we have $\frac{1}{2}-\frac{3}{4}|-9+3|=\frac{1}{2}-\frac{3}{4}|-6|=\frac{1}{2}-\frac{3}{4}(6)$, which is -4 from the above line.

So both 3 and -9 are solutions.

Example: Solve $|x + 1| = -3$.

Solution: We can't. The absolute value of a number is always nonnegative. We will never take the absolute value and get a negative number. If we forget this and try to solve the equation as normal, we have the two equations $x + 1 = -3$ or $x + 1 = 3$. Solving $x + 1 = -3$ will lead to $x = -4$, and solving $x + 1 = 3$ yields $x = 2$.

If $x = -4$, we check to see that the left-hand side of the original equation is $|-4 + 1| = |-3| = 3$, which is not the right-hand side.

If $x = 2$, then $|2 + 1| = 3$, which is again not the right-hand side of the original equation. So this equation has no solutions.

Example: Solve $|x + 1| = 0$.

Solution: If we set $x + 1 = 0$, there is no need to write another equation, because $-(0) = 0$. So when the distance from the origin to $x + 1$ is zero, we must have that $x + 1$ itself is zero. The only solution to this equation is $x = -1$.

Finally, what if we have an absolute value on both sides? If we think about it, $|x| = |y|$ will have four possibilities:

$$x = y, \quad x = -y, \quad -x = y, \quad \text{or} \quad -x = -y.$$

The middle two equations are equivalent; we are just moving the negative from one side to the other, so $x = -y$ and $y = -x$ are asking the same question. Now look at the first and last equations. They are also asking the same question, because $-x = -y$ is true when $x = y$. If there is a double absolute value, we *follow the same procedure*: We write one equation without absolute value bars, and then write another equation where we take the opposite of either side.

Example: Solve $|x + 2| = |3x + 1|$.

Solution: Having an absolute value on both sides won't change things. We can write one equation dropping the absolute value bars, and then write another equation where we take the opposite of either side:

$$\begin{array}{rcl}
x+2=3x+1 & \text{or} & x+2=-(3x+1) \\
-x \quad -x & & x+2=-3x-1 \\
2=2x+1 & & -x \quad -x \\
-1 \quad -1 & & 2=-4x-1 \\
1=2x & & +1 \quad +1 \\
\frac{1}{2}=x & \textit{or} & 3=-4x.
\end{array}$$

Our solutions are $x=\frac{1}{2}$ or $x=-\frac{3}{4}$. Checking $x=\frac{1}{2}$, we have

$$\begin{aligned}
\left|\frac{1}{2}+2\right| &= \left|3\left(\frac{1}{2}\right)+1\right| \\
\left|\frac{5}{2}\right| &= \left|\frac{3}{2}+1\right| \\
\frac{5}{2} &= \frac{5}{2}.
\end{aligned}$$

So $x=\frac{1}{2}$ works. Checking $x=-\frac{3}{4}$ yields

$$\begin{aligned}
\left|-\frac{3}{4}+2\right| &= \left|3\left(-\frac{3}{4}\right)+1\right| \\
\left|-\frac{3}{4}+\frac{8}{4}\right| &= \left|-\frac{9}{4}+1\right| \\
\left|\frac{5}{4}\right| &= \left|-\frac{9}{4}+\frac{4}{4}\right| \\
\frac{5}{4} &= \left|-\frac{5}{4}\right| \\
\frac{5}{4} &= \frac{5}{4}.
\end{aligned}$$

Both solutions, $x=\frac{1}{2}$ and $x=-\frac{3}{4}$, work.

Section 2.3 Exercises

1. Solve $|2-x|=4$.
2. Solve $|x+5|-1=1$.
3. Solve $|x-4|+1=-1$.
4. Solve $|x-4|+1=1$.
5. Solve $|-x|=2$.
6. Solve $|x|=-2$.
7. Solve $\frac{3}{4}-\frac{5}{9}|t|=1$.
8. Solve $|3x+2|=4$.
9. Solve $|3x|-5=\frac{1}{4}$.
10. Solve $|w-1|-5=3$.
11. Solve $\frac{1}{2}|3-x|=7$.
12. Solve $\frac{3}{2}|p+1|=2$.
13. Solve $\frac{1}{2}-5|v-1|=\frac{3}{2}$.
14. Solve $3-4|3x+2|=7$.
15. Solve $3+4|3x-2|=7$.
16. Solve $\frac{1}{2}+2|5-x|=1$.
17. Solve $3-|-x|=0$.
18. Solve $3-|-x|=3$.
19. Solve $3+|x-1|=3$.
20. Solve $6+\frac{1}{2}|3x-5|=6$.
21. Solve $|3x+1|=|x-5|$.
22. Solve $|2-x|=\frac{1}{2}|x-1|$.
23. Translate the statement "The distance between y and the origin is 4" into a mathematical equation. Then solve it.
24. Translate the statement "The distance between z and the origin is $\frac{1}{2}$" into a mathematical equation. Then solve it.

2.4: Applications of Single-Variable Equations

One of the most important skills in using mathematics is translating statements into a mathematical language so we can use the techniques we've learned to solve these problems.

And mathematics can feel like a new language. Luckily, with some practice and a dictionary, we can succeed.

First, if we say "is," that means =.

Ex) "Two is one plus one" translates into $2 = 1 + 1$.

Ex) "x is exactly 4 more than y" translates into $x = 4 + y$.

Another important expression is saying "of" to mean multiplication.

Ex) "One-half of four is two" translates into $\frac{1}{2} \cdot 4 = 2$.

Ex) If three-sevenths of a group of students goes to a theme park, and there are 210 students, that means that $\frac{3}{7}(210)$, or 90, of the students go.

Percent

Percents are often used to describe parts of a whole. A percent divides an amount into 100 equal parts, regardless of what the number of parts is. So $1\% = 0.01$ is the conversion that is often used.

To convert from a decimal to a percentage, move the decimal place two units to the *right*.

Ex) 0.24 is 24%.

Ex) $\frac{3}{5}$ written as a decimal is 0.6, which is 60%.

Ex) 0.001 is 0.1%.

To convert from a percentage to a decimal, move the decimal places two units to the *left*. We cannot multiply a percent by a number; the percent must be converted into decimal form.

Example: If 40% of the 800 people attending a concert said that they would come back for another show, then how many people can the venue expect to see return?

Solution: Since *of* means multiplication, we need to find 40% of 800. But we cannot multiply 40% and 800 because their units don't match, so we need to change to a proportion, or decimal. Since $40\% = 0.40$, we multiply $0.4(800)$ to get 320, which means the venue can expect 320 people to return. This makes sense, because 40% is a little less than half, and 320 is a little less than half.

Example: If there are 48 lions in a particular refuge and 25 of them are tagged, what percentage of the lions are tagged?

Solution: We know that 25 out of the 48 are tagged. The key phrase is "out of," which indicates division. So $\frac{25}{48} = 0.5208\overline{3}$ is the proportion of lions that are tagged. If we move the decimal place two units over, we get $52.08\overline{3}$, and we can round this to show that 52% of the lions are tagged. This is saying that just over half the lions are tagged, which makes sense since 25 is a little over half of 48.

Problem-Solving

Problem-solving is perhaps the most important application of mathematics. Maybe none of the following situations will apply to you in your work after college. But being able to think clearly about problems you encounter and then organize information that leads to a solution is something everyone needs to be skilled at. We present the following guidelines to assist you in solving problems.

1. What the question is asking?
2. Define helpful variables for the problem.
3. Use these variables to set up an equation.
4. Do a reality check: Does the answer make sense?
5. After solving the equation, *answer the question.*

The last step is crucial. If our boss asked us how many reams of each type of paper need to be stocked, and we answered "$x = 12$," we would probably get a frown (or worse) in response.

Example: The local bookstore charges $30 per year for a membership. The benefit is that every purchase you make during the year receives a 10% discount. If you shop there only occasionally, is the membership worth the cost? How much would you need to spend at the bookstore to break even?

Solution: We want to approach this in terms of the five guidelines presented above. (1) The number we need to find is how much we need to spend at the bookstore to recover our $30 membership fee and break even. (2) Let's define x as the amount we have to spend to break even. (3) Our equation takes 10% of x and sets it equal to 30. That is, we have

$$0.1x = 30.$$

Dividing both sides by 0.1 yields $x = 300$. So we need to spend $300 at the bookstore to recover our $30 membership. (4) This makes sense because 30 is 10% of 300. (5) If we shop there only occasionally, the membership would not be worth it. Think about it: If we spend a total of $250 throughout the year, we will receive only 10% off, or $25 off. If we pay $30, we are losing money.

Example: Suppose we need 25 liters of a 20% acid solution. Unfortunately, what we have is a liquid that is 10% acid and a liquid that is 30% acid. However, we can mix them together to create the required solution. How much of the liquids should we mix together?

Solution: Following our guidelines, (1) the question is, How much of each liquid should we add together? (2) Let's make those our variables: Let x be the amount of the 10% acid solution we should add together. If we can find x, we'll have half the answer to the question. We're asking how much of each we should add, so if we are adding x liters to another amount of the 30% solution, we want to end up with 25 liters. Put another way,

$$x + \text{amount of the other solution} = 25 \text{ liters.}$$

So the amount of the other solution must be $25 - x$. Now to solve the problem, we are worried about the concentration of acid in our solution. So (3) the equation that will solve the problem deals with the concentration of acid in our mixture.

Now, 10% of our x liters will be acid. And we are adding that to $25 - x$ liters of a 30% solution. This must equal the 25 liters being 20% acid. So these percentages get changed to decimals, and we translate "of" into multiplication. We have our equation:

$$0.1(x) + 0.3(25 - x) = 0.2(25).$$

Remember to distribute the 0.3:

$$\begin{aligned} 0.1x + 0.3(25) - 0.3x &= 5 \\ 0.1x + 7.5 - 0.3x &= 5 \\ -0.2x + 7.5 &= 5 \\ -7.5 \quad & \quad -7.5 \\ -0.2x &= -2.5 \\ \frac{-0.2x}{-0.2} &= \frac{-2.5}{-0.2} \\ x &= 12.5. \end{aligned}$$

(4) Now for the reality check: Note that 12.5 is half of 25 liters. That means $25 - x = 25 - 12.5 = 12.5$ is the amount of the 30% solution. It absolutely makes sense that half the solution should be the 10%, and the other half will be the 30% solution. We had some negative numbers up until the last step; if we had ended up with a negative number (or a number larger than 25, for that matter), we would know we made a mistake.

(5) Again, if our answer to this question is $x = 12.5$, that is complete nonsense. The question is how much of each solution should be added, so our concluding statement is this: We should add 12.5 liters of the 10% solution to 12.5 liters of the 30% solution to obtain a 25-liter solution that is 20% acid.

Example: Angela covered 30 miles biking on Orange Street and on Grove Street. She was able to pedal 3 miles per hour faster on Grove than she pedaled on Orange. Angela biked 1 hour on Orange and 2 hours on Grove. How fast was she going on each street?

Solution: (1) The question is, How fast was Angela going on Orange, and how fast was she going on Grove? (2) Let's pick one of these—Orange or Grove—as our variable. Let x equal Angela's speed on Orange Street. Then what do we know about her speed on Grove Street? It is 3 mph *faster*. So Angela's speed on Grove was $x + 3$.

We will need the "distance equals rate times time" formula to help us. Angela's total distance was 30 miles, split between Orange and Grove. Let's add together how far Angela went on Orange and how far she went on Grove.

Angela's distance on Orange is

$$(rate)(time) = x \cdot 1,$$

since she traveled x mph on Orange for 1 hour. Angela's distance on Grove is

$$(rate)(time) = (x+3) \cdot 2,$$

since she traveled $x + 3$ mph on Grove for 2 hours. Her total distance (30 miles) is the sum of these two quantities. (3) So we have

$$x + 2(x + 3) = 30.$$

Distributing the 2 and then solving gives us

$$\begin{aligned} x + 2x + 6 &= 30 \\ 3x + 6 &= 30 \\ 3x &= 24 \\ x &= 8. \end{aligned}$$

(4) This seems about right: $8 \cdot 1 = 8$ miles traveled on Orange and $2(8 + 3) = 22$ miles traveled on Grove totals up to 30 miles.

(5) Again, we must be careful not to answer this question with $x = 8$. We are looking for Angela's speed on each street, which was x and $x + 3$. So Angela traveled 8 mph on Orange and 11 mph on Grove.

Section 2.4 Exercises

1. How many pounds of Brazilian coffee beans that cost \$4 per pound should be added to 7 pounds of Jamaican coffee beans that cost \$9 per pound to produce a blend that costs \$6 per pound?

2. A lawn-mowing company charges $18 per cut to maintain a lawn. If a new lawn mower costs $300 dollars, how many lawn cuts add up to the cost of the lawn mower?
3. Suppose it takes you 45 minutes to cut the lawn. If you pay a company to take care of the lawn 12 times, how many hours are you saving by not cutting the lawn those 12 times?
4. Suppose you want to put in a garden on the side of your house, and you need the garden's length to be twice its width. If the total perimeter of the garden needs to be 40 feet, what should the dimensions of the garden be?
5. What number is two-fifths of 150?
6. 150 is two-fifths of what number?
7. Only 350 of the 800 students were able to attend the meeting. What percentage of students attended?
8. At a faculty meeting, 62 out of 157 faculty members attended. If there needs to be a quorum of 45% of all teachers present in order to make a decision, were there enough faculty members at the meeting to reach a quorum?
9. A bag of candy contains red- and blue-colored candies. If there are twice as many blue candies as red candies in each bag and every bag contains precisely 126 candies, how many of each color of candy is in a bag?
10. Only hummingbirds and cardinals are found in a particular area of the forest. If 35% of the birds are hummingbirds, and there are a total of 150 birds in the area, how many cardinals are in the area?
11. Suppose you need 30 liters of antifreeze in your radiator. Your car requires a solution that is 55% antifreeze. If you have a solution that is 70% antifreeze and another solution that is 30% antifreeze, how much should of each solution should you mix together for the desired result?
12. Two different blends of tea are mixed to create a new blend. The new blend is 2 pounds of 35% oolong tea. If the first blend of tea was 15% oolong, and the second blend of tea was 60% oolong, how many pounds of each blend was mixed for the new blend?
13. A butcher is mixing two types of ground beef: one that is 30% fat and a leaner one that is 7% fat. The butcher is trying to get 20 pounds of a mixture that is 15% fat. How much of each type of ground beef should be used? Round to the nearest hundredth.
14. Kristin drove for 6 hours on a road trip to the Panhandle. For the first half of her trip she traveled 2 hours on I-75, and the rest of the time she traveled on I-10. Because of construction, her speed on I-75 was 15 mph slower than her speed on I-10. How fast did Kristin travel on I-10?

15. Lauren likes to alternate between running and walking. She can run at 6 mph and she walks at 2 mph. If Lauren spent two hours either running or walking, how much time was she running and how much time was she walking?
16. Critical thinking: Write down a problem about a mixture of acid that corresponds to the following equation: $0.14x + 0.58(36 - x) = 0.3(36)$.

Chapter 2 Review

2-A

1. We need a 40-liter solution that is 25% acid. We have available to us as much as we need of a 10% acid solution and a 40% acid solution. How much of each type of solution should be mixed to obtain the desired result?
2. Solve $x + 4 = -3$
3. Solve $3(x - 5) = 2(6 - x)$.
4. Solve the equation $A = \frac{1}{2}bh$ for h.
5. Solve the equation $|x - 5| = 4$.
6. Solve the equation $|5 - w| = |2w + 1|$ for w.
7. If one-fifth of a country's population is children, what percentage of that population is children?
8. What is 30% of 180?

2-B

1. 45% of what number is 300?
2. Solve the equation $|3x - 5| = |4(x - 1) + 2|$.
3. Solve the equation $\frac{1}{2}(5 - 3x) = \frac{3}{4}(2x + 1)$.
4. Solve the equation

$$3a(b + d) = 4(c - d)$$

for d.

5. Solve the equation $|2(x-1)-5|-3=0$.
6. Roughly three-quarters of the blue jays in the forest are aggressive. If there are 500 blue jays in the forest, how many are aggressive?
7. Solve $2y-5+3y=5(1-y)$.
8. A new blend of tea that is 40% rooibos and weighs 6 pounds is created by mixing two other blends of red tea. If the first blend was 18% rooibos, and the second blend was 55% rooibos, how much of each blend was added together to create the final blend?

CHAPTER

3

Lines, Equations, and Systems

The previous chapter dealt mostly with equations in one variable. What's more interesting is when another variable depends on the first variable. For instance, in this chapter we will discuss equations that involve both x and y. The convention is usually that x is the independent variable and y is the dependent variable. This is because we will write y in terms of x. And if we know the value of the independent variable x, we can figure out the value of y by plugging the x-value into some equation.

But we can use any variables we want. For instance, if we know the stock price p depends linearly on time t, then t is our independent variable and p is the dependent variable.

3.1: Lines and Their Intercepts

The xy-plane, or Cartesian plane, is made up of ordered pairs of real numbers (x, y). There are two axes in the plane: the x-axis and the y-axis. The x-axis is horizontal, while the y-axis is vertical. We can think of the coordinate axes as two number lines that cross at right angles. The coordinate axes split the plane into four quadrants. See Figure 3-1.

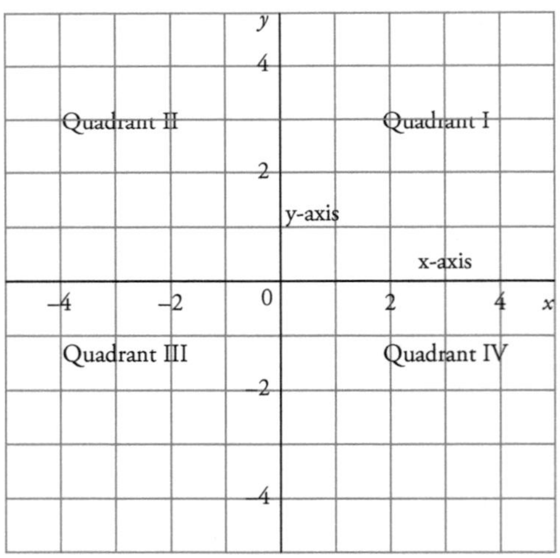

Fig. 3-1

Points in the coordinate plane move according to the number lines. The point (1, 2) is a point that is 1 unit away from the origin in the positive x direction, and 2 units away from the origin in the y direction.

The origin is the point (0, 0). Points in the coordinate plane move according to the number lines. The point (1, 2) is a point that is 1 point away from the origin in the positive x direction, and 2 point away from the origin in the y direction.

The first quadrant, or Quadrant I, has points in it where both the x- and y-coordinates are positive.

Ex) The points (1, 2) and (2, 4) are plotted in the first quadrant. See Figure 3-2.

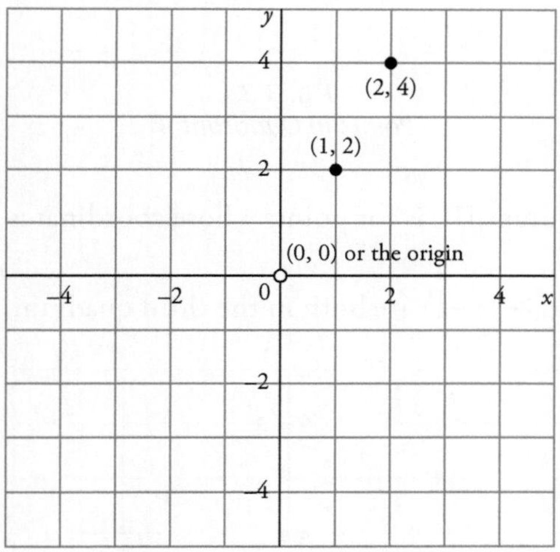

Fig. 3-2
Points in Quadrant I

The second quadrant, or Quadrant II, is for points whose x-coordinates are negative and y-coordinates are positive.

Ex) The points $(-2, 3)$ and $\left(-\frac{3}{2},1\right)$ are both in the second quadrant. See Figure 3-3.

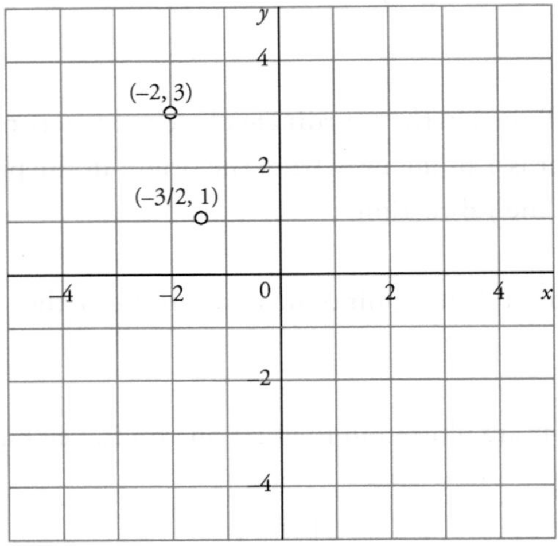

Fig. 3-3
Points in Quadrant II

The <u>third quadrant</u>, or Quadrant III, is for points whose coordinates are both negative.

Ex) The point $\left(-\sqrt{3},-5\right)$ and $(-1,-1)$ are both in the third quadrant. See Figure 3-4.

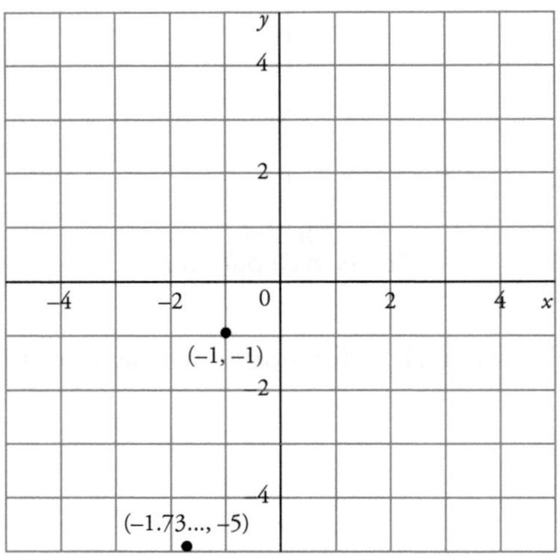

Fig. 3-4
Points in Quadrant III

The fourth quadrant, or Quadrant IV, is for ordered pairs with positive x-coordinates and negative y-coordinates.

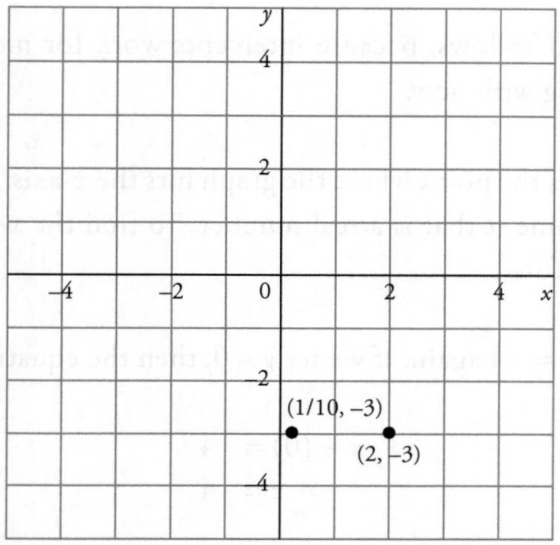

Fig. 3-5
Points in Quadrant IV

Ex) The points $(2, -3)$ and $\left(\frac{1}{10}, -3\right)$ are points in the fourth quadrant. See Figure 3-5.

The first step in learning to graph equations is to start with lines.

The first-order linear equation in two variables is

$$Ax + By + C = 0$$

where A, B, and C are real numbers and x and y are variables. The graph of an equation is the set of ordered pairs (x, y) that satisfy the equation.

Ex) Consider the equation $x - y = -4$. Note that the ordered pair $(1, 5)$ satisfies this equation, because if $x = 1, y = 5$, then $x - y = (1) - (5) = -4$. Thus, the point $(1, 5)$ lies on the graph of $x - y = -4$. If we can find more points that satisfy the equation, we can plot those points and connect the dots to form the graph.

Recall from geometry that we need only two points to determine a line. The two points that are used most often to graph a line are the intercepts of the line.

Pay close attention to what follows, because intercepts work for more complicated equations than the kind we are dealing with now.

The x-intercept of a graph is the point where the graph hits the x-axis. Points on the x-axis always have the form $(\alpha, 0)$ for some α that is a real number. To find the x-intercept of a line, we just need to set $y = 0$ and solve for x.

Ex) Consider the line $x - y = -4$ again. If we set $y = 0$, then the equation we end up with is

$$\begin{aligned} x - (0) &= -4 \\ x &= -4 \end{aligned}$$

The line will hit the x-axis at the point $(-4, 0)$, and we know that $(-4, 0)$ is a point on the line.

Similarly, the y-intercept of a graph is the point where it hits the y-axis. Any point on the y-axis has the form $(0, \beta)$ where β is a real number. So to find the y-intercept, we set $x = 0$ and solve for y.

Ex) If we set $x = 0$ in our line $x - y = -4$, we have

$$\begin{aligned} (0) - y &= -4 \\ -y &= -4 \\ y &= 4 \end{aligned}$$

Thus, the point $(0, 4)$ lies on our line. It is the y-intercept.

Now that we have the two intercepts, we can graph our line by plotting the points (–4, 0) and (0, 4) and connecting those two points. See Figure 3-6.

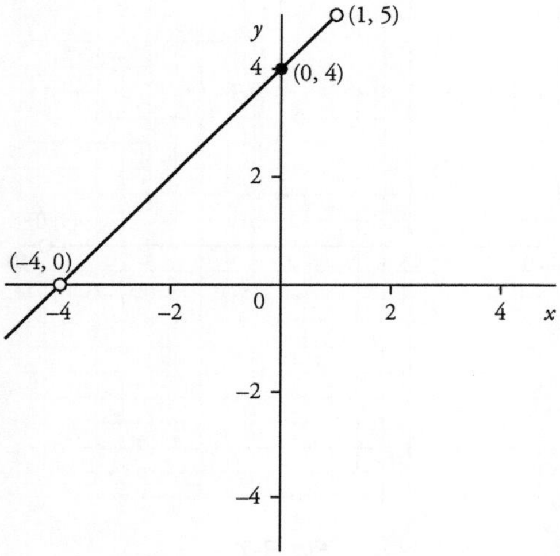

Fig. 3-6
Graph of $x - y = -4$

Example: Find the intercepts of $2x + 3y = 18$, and graph the line.

Solution: To find the x-intercept, let $y = 0$. We have

$$\begin{aligned} 2x + 3(0) &= 18 \\ 2x &= 18 \\ \frac{1}{2}(2x) &= \frac{1}{2}(18) \\ x &= 9. \end{aligned}$$

Remember to write the x-intercept as an ordered pair: (9, 0). To find the y-intercept, let $x = 0$. We have

$$\begin{aligned} 2(0) + 3y &= 18 \\ 3y &= 18 \\ \frac{3y}{3} &= \frac{18}{3} \\ y &= 6. \end{aligned}$$

So the y-intercept is $(0, 6)$. First we plot the intercepts in Figure 3-7.

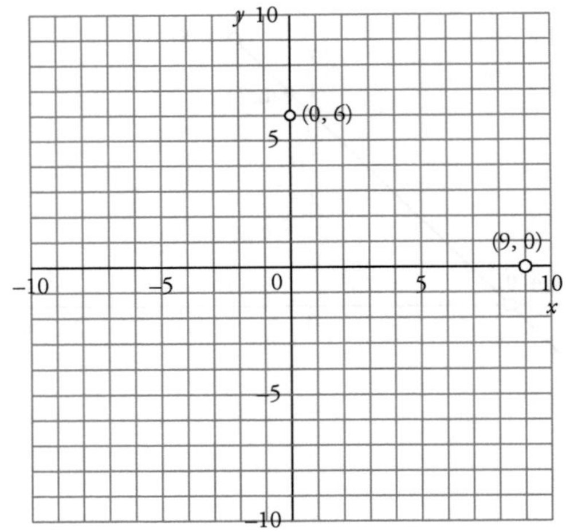

Fig. 3-7

Then we can connect the points to graph the line as in Figure 3-8.

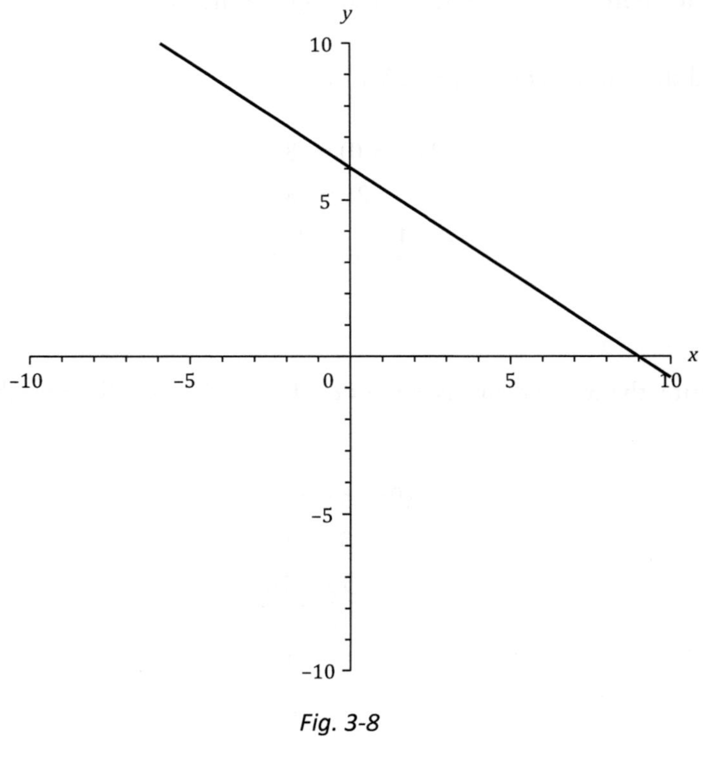

Fig. 3-8

T-Charts

Another way we can graph a line is to solve for y in terms of x. Then, if we pick different values for x, we can plug them in for the corresponding y's. These will make ordered pairs that will lie on the line.

Example: Graph the line $y = 4x - 5$.

Solution: If we make a chart, we will pick x's in the first column and find the corresponding y's in the second column. This is called "making a t-chart." We have these values:

x	$y = 4x - 5$
0	$y = 4(0) - 5 = -5$
1	$y = 4(1) - 5 = -1$
2	$y = 4(2) - 5 = 8 - 5 = 3$

So the ordered pairs $(0, -5)$, $(1, -1)$, and $(2, 3)$ all lie on the graph of $y = 4x - 5$.

Section 3.1 Exercises

1. Plot the points $\left(\frac{1}{2}, 3\right)$, $(-2, 1)$, $(4, 4)$, and $(4, -1)$ on the same set of axes.
2. a. In which quadrant is the point $(\pi, 2)$?
   b. In which quadrant is the point $(-6, \pi)$?
   c. In which quadrant is the point $(\pi - 5, 1)$?
   d. In which quadrant is the point $(2, -\pi)$?
   e. In which quadrant is the point $(-\pi, 0)$?
3. Does the point $(-1, 3)$ lie on the line $2x + y = 6$? Why or why not?
4. Does the point $(-10, 26)$ lie on the line $2x + y = 6$? Why or why not?
5. Find the intercepts of $2x + 5y = 10$ and graph the line.
6. Find the intercepts of $3x + 4y = -12$ and graph the line.
7. Find the intercepts of $5x - 6y = 30$ and graph the line.
8. Find the intercepts of $-2x + 9y = 90$ and graph the line.
9. Find the intercepts of $3x - 7y = 21$ and graph the line.
10. Find the intercepts of $-x - y = -1$ and graph the line.
11. Find the intercepts of $2x + 4y = 5$ and graph the line.

12. Find the intercepts of $x - 2y = 3$ and graph the line.
13. Find the intercepts of $-3x + 5y = 10$ and graph the line.
14. Find the intercepts of $\frac{1}{2}x + \frac{1}{4}y = 1$ and graph the line.
15. Find the intercepts of $\frac{1}{3}x - \frac{2}{3}y = 3$ and graph the line.
16. Looking ahead: Consider the equation $y = x^2 + 2x - 2$. We will learn how to graph something like this in Chapter 8. For now, re-read the definition of y-intercept. Find the ordered pair that is the y-intercept of $y = x^2 + 2x - 2$.
17. Looking ahead: Consider the equation $y = x^2 + 2x - 2$. Make a t-chart using several x-values, like $-3, -2, -1, 0, 1, 2$, and 3. Then sketch the graph by plotting the points and connecting them.

3.2: Lines and Their Slopes

The slope of a line is defined as

$$\text{slope} = \frac{\text{rise}}{\text{run}}.$$

For example, if a line has a slope of 2, we can think of 2 as $\frac{2}{1}$. So for every rise of 2 units in the positive y direction, there is a run of 1 unit in the positive x direction.

Ex) In Figure 3-9 we have a line with a slope of 2. For every rise of 2 units in the y direction, the line moves 1 unit in the x direction.

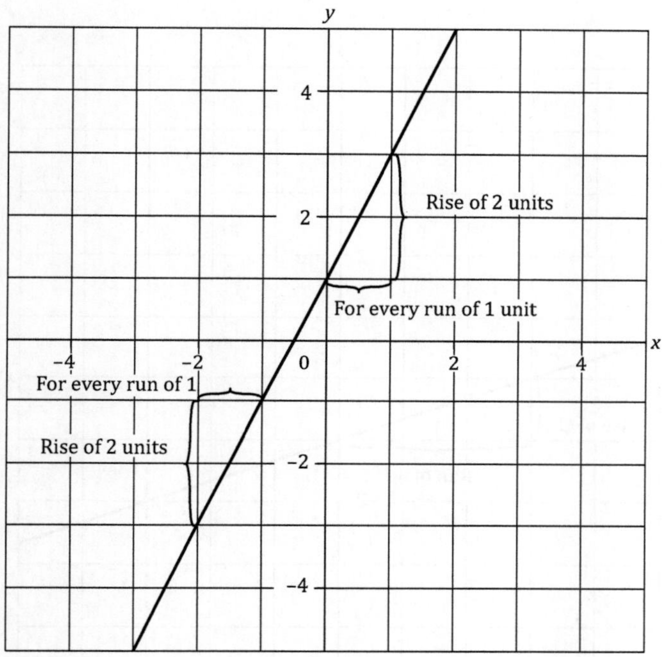

Fig. 3-9

Ex) In Figure 3-10 we have a line with a slope of $-\frac{1}{3}$. We would think of $-\frac{1}{3}$ as $\frac{-1}{3}$. So this line has a rise of -1 for every run of 3. So we move downward (the negative y direction) 1 unit for every 3 units of run (in the positive x direction).

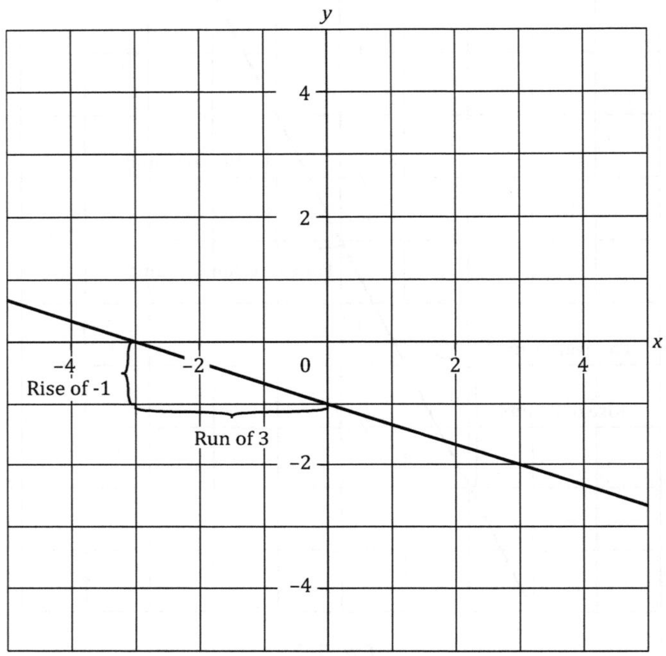

Fig. 3-10

Now, we can use any two points on a line to determine its slope. In general, consider the points (x_1, y_1) and (x_2, y_2) to be points on the same line. See Figure 3-11.

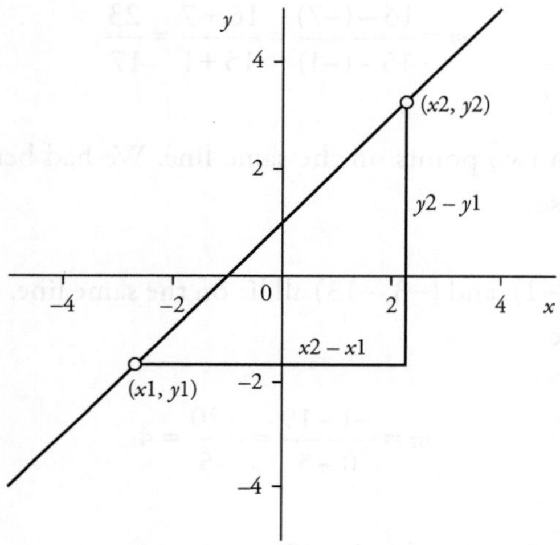

Fig. 3-11

The rise of the line is $(y_2 - y_1)$ for every run of $(x_2 - x_1)$. We are ready for a formula. The slope of the line through (x_1, y_1) and (x_2, y_2) is

$$\boxed{m = \frac{y_2 - y_1}{x_2 - x_1}.}$$

Example: Find the slope of the line through (4, 6) and (−1, 2).

Solution: The point (4, 6) will be our (x_1, y_1), and the point (−1, 2) will be our (x_2, y_2). We take the difference in the *y*'s divided by the difference in the *x*'s. We have

$$m = \frac{2-6}{-1-4} = \frac{-4}{-5} = \frac{4}{5}.$$

Example: Find the slope of the line through (−1, −7) and (15, 16).

Solution: Take the difference in the y-values over the difference in the x-values. Be careful when subtracting negative numbers; it usually helps to put them in parentheses.

$$m = \frac{16-(-7)}{15-(-1)} = \frac{16+7}{15+1} = \frac{23}{17}.$$

Suppose we had more than two points on the same line. We had better get the same slope no matter which two we choose.

Ex) The points (5, 19), (0, −1), and (−3, −13) all lie on the same line. If we use the pair of points (5, 19) and (0, −1), we have

$$m = \frac{-1-19}{0-5} = \frac{-20}{-5} = 4.$$

If we use the pair of points (0, −1) and (−3, −13), we have

$$m = \frac{-13-(-1)}{-3-0} = \frac{-13+1}{-3} = \frac{-12}{-3} = 4.$$

YOU TRY IT: Find the slope of the previous line using the remaining pair (5, 19) and (−3, −13), and show that it is the same slope as above.

Example: Find the slope of the line between (4, 7) and (−15,7).

Solution: The slope is

$$m = \frac{7-7}{-15-4} = \frac{0}{-19} = 0.$$

The slope is zero. In this case, the line is horizontal. See Figure 3-12.

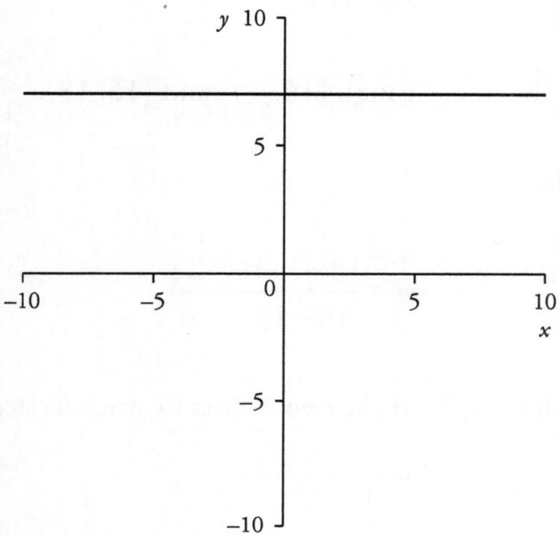

Fig. 3-12
Graph of the horizontal line y = 7.

Now, there are also vertical lines. How should we define their slope? See Figure 3-13.

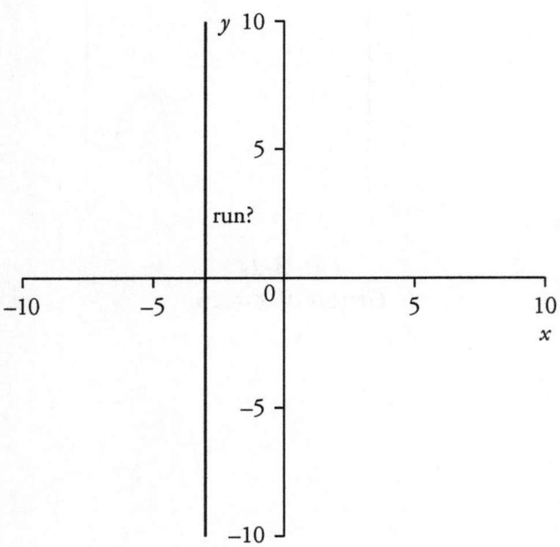

Fig. 3-13
Given any rise, a vertical line has no run.

What rise do they have? There is no run. We say that the slope of these lines is undefined (because it is infinite).

Example: Find the slope of the line through (16, –4) and (16, 18).

Solution: Using our formula,

$$m = \frac{18-(-4)}{16-16} = \frac{22}{0},$$

so this slope is undefined. That means these two points lie on a vertical line. See Figure 3-14.

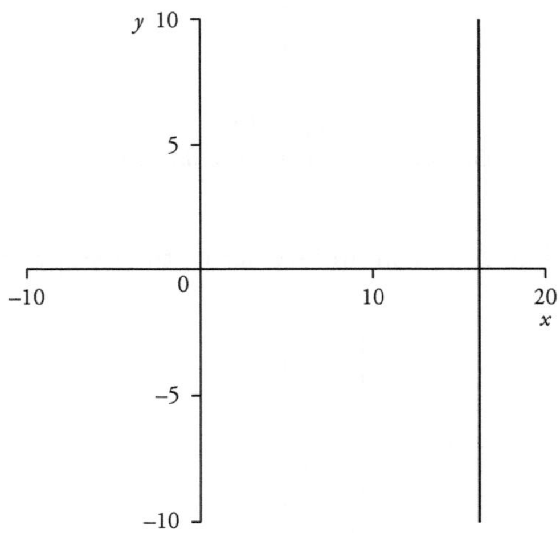

Fig. 3-14
Graph of x = 16.

Parallel and Perpendicular Lines

Two lines are <u>parallel</u> if they do not intersect. See Figure 3-15.

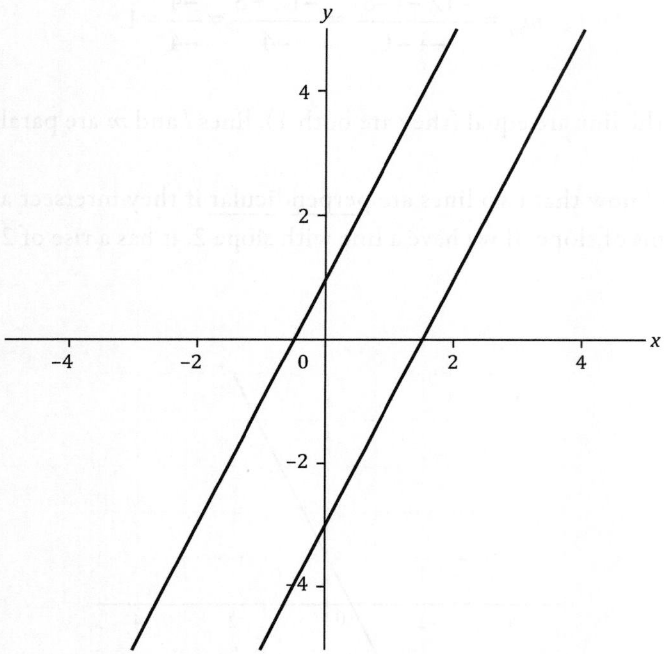

Fig. 3-15
Parallel lines do not intersect.

It should make sense that *parallel lines have the same slope.*

Example: The points (1, 2) and (5, 6) lie on line l, and the points (0, −8) and (−4, −12) lie on line m. Show that lines l and m are parallel.

Solution: Find the slope of line l. We will use $x_1 = 1, y_1 = 2, x_2 = 5, y_2 = 6$. We have

$$m_l = \frac{6-2}{5-1} = \frac{4}{4} = 1,$$

where we are using the notation m_l to denote the slope of line l. Now let's calculate the slope of line m using $x_1 = 0, y_1 = -8, x_2 = -4, y_2 = -12$:

$$m_m = \frac{-12-(-8)}{-4-0} = \frac{-12+8}{-4} = \frac{-4}{-4} = 1.$$

Since the slopes of the line are equal (they are both 1), lines l and m are parallel.

From geometry, we know that two lines are perpendicular if they intersect at right angles. Let's consider this in terms of slope. If we have a line with slope 2, it has a rise of 2 and a run of 1. See Figure 3-16.

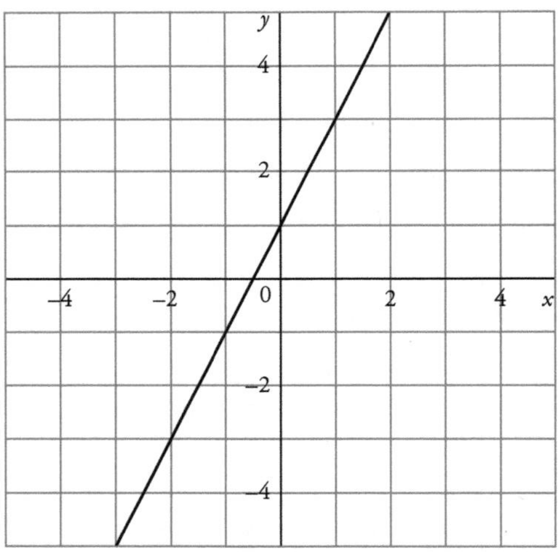

Fig. 3-16
Line with slope of 2.

To make another line intersect this one at a right angle, our new line should have a rise of -1 and run of 2. See Figure 3-17.

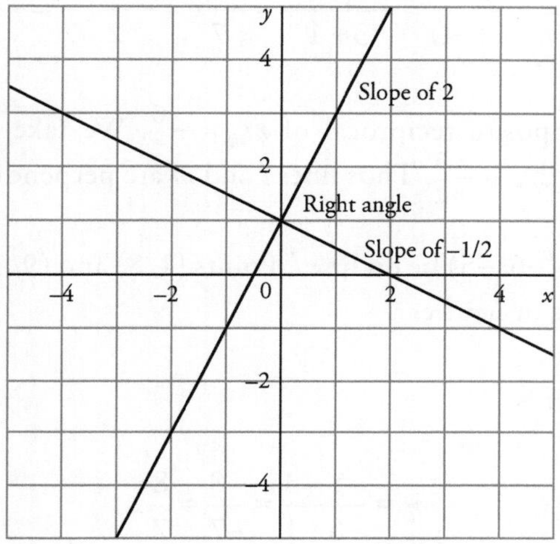

Fig. 3-17
Perpendicular Lines

Then we can say that *perpendicular lines have opposite reciprocal slopes*. That is, if line l has a slope of $\frac{a}{b}$, then any line perpendicular to l will have slope $-\frac{b}{a}$.

Ex) If line l has a slope of 4, then any line perpendicular to line l will have a slope of $-\frac{1}{4}$.

Ex) If line m has a slope of $\frac{2}{3}$, then any line perpendicular to line m will have a slope of $-\frac{3}{2}$.

Example: Line l passes through $\left(1, \frac{14}{3}\right)$ and $(-3, 2)$. Line m passes through $(-4, 1)$ and $\left(3, -\frac{19}{2}\right)$. Show that lines l and m are perpendicular.

Solution: Let's find the slope of line l. We have

$$m_l = \frac{2-\frac{14}{3}}{-3-1} = \frac{\frac{6}{3}-\frac{14}{3}}{-4} = \frac{-\frac{8}{3}}{-4} = \frac{8}{3} \div 4 = \frac{8}{3}\cdot\frac{1}{4} = \frac{2}{3}.$$

The slope of line m is

$$m_m = \frac{-\frac{19}{2}-1}{3-(-4)} = \frac{-\frac{19}{2}-\frac{2}{2}}{3+4} = \frac{-\frac{21}{2}}{7} = -\frac{21}{2}\cdot\frac{1}{7} = -\frac{3}{2}.$$

Note that $m_l = \frac{2}{3}$ is the opposite reciprocal of $m_m = -\frac{3}{2}$. We take the opposite of $\frac{2}{3}$, which is $-\frac{2}{3}$, and the reciprocal of this is $-\frac{3}{2}$. Thus lines l and m are perpendicular.

Example: Points (1, 3) and (−6, −5) lie on line l. Points (1, 8) and (9, 15) lie on line m. Are these lines parallel, perpendicular, or neither?

Solution: The slope of line l is

$$m_l = \frac{-5-3}{-6-1} = \frac{-8}{-7} = \frac{8}{7}.$$

The slope of line m is

$$m_m = \frac{15-8}{9-1} = \frac{7}{8}.$$

Note that $\frac{8}{7}$ and $\frac{7}{8}$ are just reciprocals. So these lines are neither parallel nor perpendicular. The slopes would have to be equal reciprocals, or opposite reciprocals, respectively.

Example: Line l passes through points (6, −3) and (−2, 5). What is the slope of any line parallel to l? What is the slope of any line perpendicular to l?

Solution: The slope of line l is

$$m = \frac{5-(-3)}{-2-6} = \frac{5+8}{-8} = \frac{13}{-8} = -\frac{13}{8}.$$

So the slope of any line parallel to l is also $-\frac{13}{8}$, and the slope of any line perpendicular to l is $\frac{8}{13}$.

Section 3.2 Exercises

1. Find the slope of the line through (1, 2) and (3, 4).
2. Find the slope of the line through (−1, 6) and (14, −7).
3. Find the slope of the line through (2, −5) and (20, −10).
4. Find the slope of the line through (−1, −5) and (16, −6).
5. Line l contains points (4, −3) and (0, 2). Line m contains points (25, 16) and (17, 26). Are lines l and m parallel, perpendicular, or neither?
6. Line m contains points (14, 27) and (−60, 25). Lines parallel to line m will have what slope? Lines perpendicular to m will have what slope?
7. Find the x- and y-intercepts of $4x + 10y = 60$. Use the intercepts to find the slope of the line.
8. Find the slope of the line through (14, 30) and (−600, 30). Graph the line.
9. Find the slope of the line through (−6, 10) and (−6, 15). Graph the line.
10. Find the slope of the line through (614, 3) and $(\sqrt{3},3)$.
11. Find the slope of the line through (π, 15) and (π, −6).
12. Line l contains points (1, 10) and (−7, −30). Line m contains points (5, −1) and $\left(-6,-\frac{16}{5}\right)$. Line n contains points (20, 2) and (−5, 7). What relationship do these lines have? Are they parallel to each other, perpendicular, or neither?
13. Think about it: The slope of line could also be defined as $\frac{y_1-y_2}{x_1-x_2}$. Show that this is the same as the slope formula by multiplying $\frac{y_1-y_2}{x_1-x_2}$ by $\frac{-1}{-1}$.
14. Think about it: Two lines are perpendicular if their slopes are opposite reciprocals of each other. Show that this is the same thing as saying that the product of their slopes is −1. (Hint: Let m_1 and m_2 be the slopes. What does it mean for m_1 and m_2 to be opposite reciprocals?)

3.3: LINES AND THEIR EQUATIONS

Given information about a line, we would like to be able to write its equation. In this section, we will discuss the two ways in which this is done: using point-slope form, and using slope-intercept form. They are aptly named.

The equation of the line with slope m passing through (x_1, y_1) is

$$\boxed{y - y_1 = m(x - x_1).}$$

This is the point-slope form of the equation of a line. Notice that we need a point (x_1, y_1) that lies on the line and the slope m of the line.

Example: Write the equation of the line through (6, 7) with slope $-\frac{3}{2}$.

Solution: Since (6, 7) is a point on the line, we have $x_1 = 6$, $y_1 = 7$. The slope of the line is $m = -\frac{3}{2}$. So the equation of the line in point-slope form is $y - 7 = -\frac{3}{2}(x - 6)$.

Example: Find the equation of the line through (−3, 4) and (6, −2).

Solution: To use point-slope form, we need a point on the line and the slope of the line. Let's start by finding the slope:

$$m = \frac{-2 - 4}{6 - (-3)} = \frac{-6}{6 + 3} = \frac{-6}{9} = -\frac{2}{3}.$$

Now, we can use any point on the line. So let's use (−3, 4) as (x_1, y_1). The equation of our line is

$$y - 4 = -\frac{2}{3}(x - (-3)),$$

or $y - 4 = -\frac{2}{3}(x + 3)$. If we distribute the $-\frac{2}{3}$, we can simplify this to

$$y - 4 = -\frac{2}{3}x - \frac{2}{3}(3),$$

which is

$$y - 4 = -\frac{2}{3}x - 2.$$

Adding 4 to both sides gives us

$$y = -\frac{2}{3}x + 2.$$

YOU TRY IT: Use the point (6, −2) in the previous example as (x_1, y_1) and show that you will get the same equation of the line (using $m = -\frac{2}{3}$).

If we take point-slope form and solve for y, then we can simplify and write the equation in slope-intercept form:

$$\boxed{y = mx + b.}$$

This is called slope-intercept form because m is the slope and the y-intercept is $(0, b)$. Slope-intercept form presents an easy way to graph an equation. We can find the spot where this line passes through on the y-axis, and then use the slope to find another point and connect them.

Example: Graph the line $y = \frac{3}{2}x - 2$.

Solution: First we plot the y-intercept, which is (0, −2). Then we use the slope, which is $\frac{3}{2}$, to get to other points. We have a rise of 3 and a run of 2. So we move up 3 units, then over 2 units, and make a dot. This will be another point on the line. We can continue in this way, making as many points as we like and connecting them to make a sketch of the graph. See Figure 3-18.

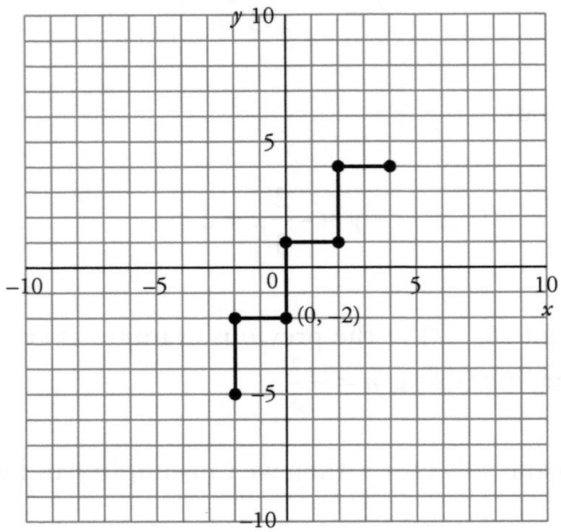

Fig. 3-18

Note that we can also go down 3 and to the left 2. This is equivalent to $\frac{-3}{-2}$, which we know is equal to $\frac{3}{2}$.

The finished graph of $y = \frac{3}{2}x - 2$ is in Figure 3-19.

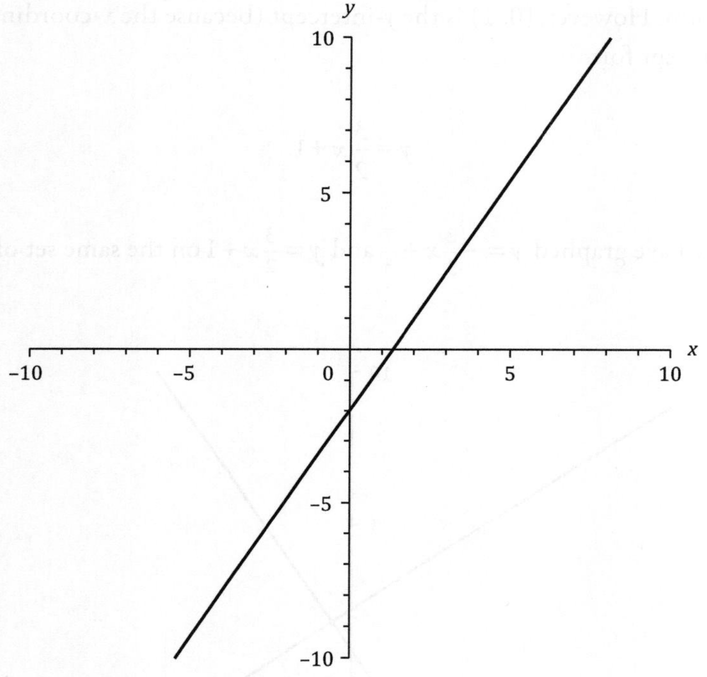

Fig. 3-19

Example: Find the equation in slope-intercept form of the line perpendicular to $2x + 3y = 7$ passing through (0, 1). Graph both lines.

Solution: We are looking for a line perpendicular to $2x + 3y = 7$, then we need a slope that is the opposite reciprocal of this line's slope. If we put the equation in slope-intercept form, we can read the slope of the line right off the equation. Solving for y, we have

$$\begin{aligned} 2x + 3y &= 7 \\ -2x \quad &\ \ -2x \\ 3y &= -2x + 7 \\ \frac{1}{3}(3y) &= \frac{1}{3}(-2x + 7) \\ y &= -\frac{2}{3}x + \frac{7}{3}. \end{aligned}$$

The slope of this line is $-\frac{2}{3}$, and the slope of the line we are looking for is $\frac{3}{2}$. Now, our new line also passes through (0, 1). If we write the equation with slope $\frac{3}{2}$ passing through (0, 1), we can use point-slope form. However, (0, 1) is the y-intercept (because the x-coordinate is zero), so we can use slope-intercept form:

$$y = \frac{3}{2}x + 1.$$

In Figure 3-20, we have graphed $y = -\frac{2}{3}x + \frac{7}{3}$ and $y = \frac{3}{2}x + 1$ on the same set of axes.

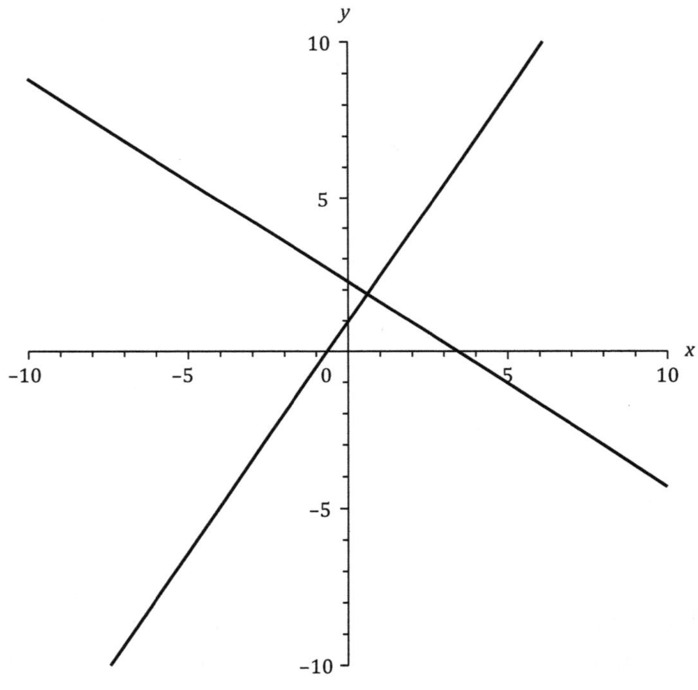

Fig. 3-20

Example: Find the equation in slope-intercept form of the line parallel to $y = 5 - x$ passing through the origin. Graph both lines.

Solution: Any line parallel to $y = -x + 5$ will have slope -1, because the slope will be equal. So we want the equation of a line with slope -1 passing through $(0, 0)$. Note that we can use point-slope form for

$$y - 0 = -1(x - 0),$$

which simplifies to $y = -x$. The graph of $y = -x + 5$ and $y = -x$ is given in Figure 3-21.

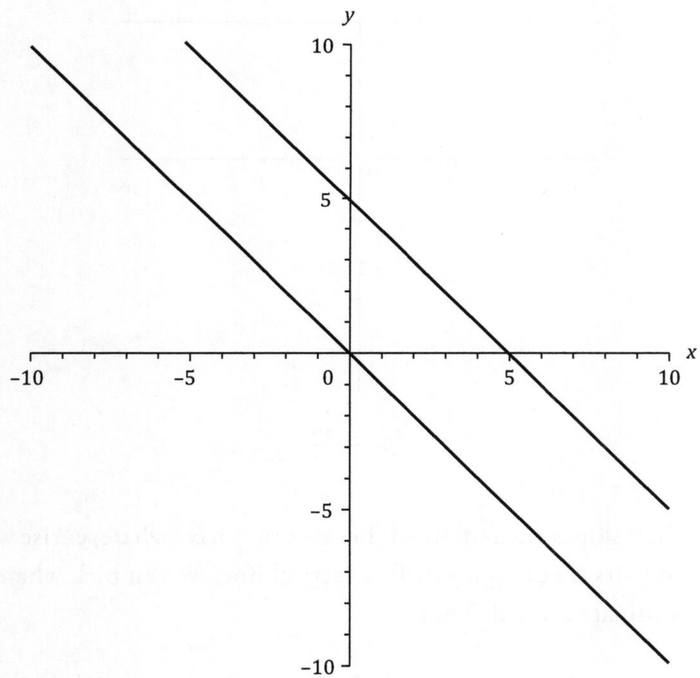

Fig. 3-21

Recall the discussion of horizontal and vertical lines in the previous section. A horizontal line has a slope of zero. So if $m = 0$ in the slope-intercept formula, we end up with $y = b$. Therefore,

any horizontal line has the form $y = b$.

Ex) The line $y = 6$ is a horizontal line. In terms of ordered pairs that satisfy the equation, note that we can plug in anything we like for x. So the points (0, 6), (15, 6), and (−130, 243, 6) all lie on the line $y = 6$. We can think of this line as $y = 0x + 6$. It has slope zero and y-intercept 6. See Figure 3-22.

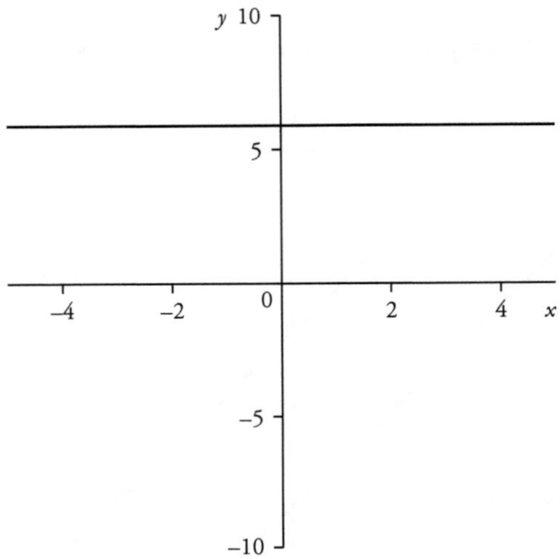

Fig. 3-22

As for vertical lines, their slopes are undefined (because they have whatever rise we want over a run of 0). For the ordered pairs that might satisfy a vertical line, we can pick whatever y-coordinate we like, but the x-coordinate is fixed. Thus,

any vertical line has the form $x = a$.

Ex) The line $x = -2$ is a vertical line that hits the x-axis at $(-2, 0)$. See Figure 3-23.

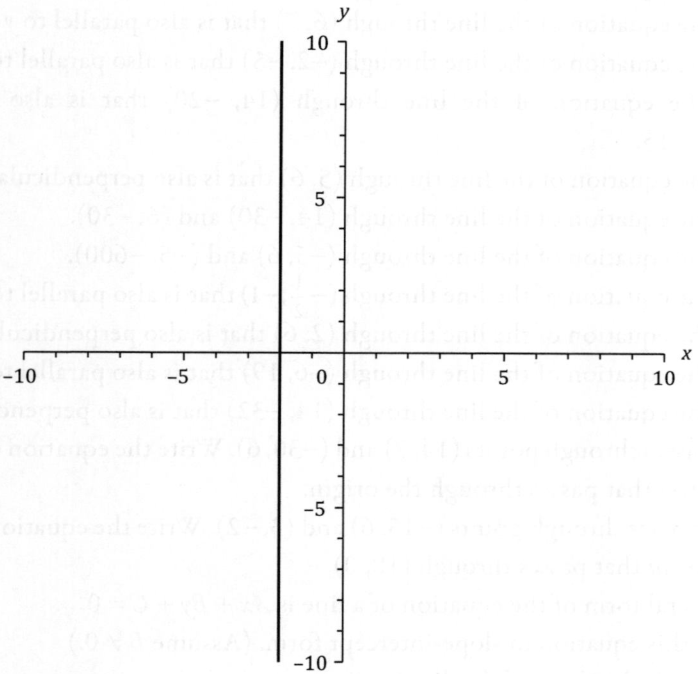

Fig. 3-23

Section 3.3 Exercises

1. Graph the line $y = -3x + 1$.
2. Graph the line $y = 4x - 10$.
3. Graph the line $2y = x + 4$.
4. Graph the line $y = -\frac{3}{5}x + 2$.
5. Graph the line $y = -\frac{7}{2}x$.
6. Put the equation $4x + 10y = 60$ in slope-intercept form and graph it.
7. Put the equation $x + 2y = 9$ in slope-intercept form and graph it.
8. Write the equation of the line that passes through $(1, 10)$ and $(-7, -30)$.
9. Write the equation of the line that passes through $(5, -1)$ and $\left(-6, -\frac{16}{5}\right)$.
10. Write the equation of the line that passes through $(20, 2)$ and $(-5, 7)$.
11. Write the equation of the line with slope 4 and y-intercept 6.
12. Write the equation of the line with slope $-\frac{1}{2}$ and x-intercept 2.

13. Write the equation of the line through $(-2, -6)$ with slope $\frac{2}{3}$ in slope-intercept form.
14. Write the equation of the line through $(16, 15)$ with slope -6 in slope-intercept form.
15. Write the equation of the line through $(6, 7)$ that is also parallel to $y = 3x + 1$.
16. Write the equation of the line through $(-2, -5)$ that is also parallel to $2x + 5y = 10$.
17. Write the equation of the line through $(14, -20)$ that is also perpendicular to $y = -5x - 15$.
18. Write the equation of the line through $(5, 6)$ that is also perpendicular to $15x + 6y = 9$.
19. Write the equation of the line through $(14, -30)$ and $(6, -30)$.
20. Write the equation of the line through $(-5, 6)$ and $(-5, -600)$.
21. Write the equation of the line through $(-\frac{1}{2}, -1)$ that is also parallel to $y = 5$.
22. Write the equation of the line through $(2, 6)$ that is also perpendicular to $y = 5$.
23. Write the equation of the line through $(-6, 19)$ that is also parallel to $x = 7$.
24. Write the equation of the line through $(14, -32)$ that is also perpendicular to $x = 10$.
25. Line l passes through points $(14, 7)$ and $(-30, 6)$. Write the equation of the line perpendicular to l that passes through the origin.
26. Line m passes through points $(-15, 6)$ and $(3, -2)$. Write the equation of the line parallel to line m that passes through $(18, 0)$.
27. The general form of the equation of a line is $Ax + By + C = 0$.
    a. Put this equation in slope-intercept form. (Assume $B \neq 0$.)
    b. What is the slope of this line?
    c. What is the y-intercept of this line?
    d. What is the x-intercept of this line?
    e. Use your answers to parts c) and d) to find the slope of the line using the intercepts, and show that this is the same thing you found in part b).

3.4: Two-Variable Systems of Linear Equations

In this section we are concerned with the intersection of two lines in the coordinate plane. This is an important application in the study of lines, and we will discuss many applications of it in Section 3.6.

If we graph two lines in the xy-plane, there are three options:

Option 1: The lines never intersect. For example, note the lines $y = 2x - 1$ and $y = 2x + 3$. See Figure 3-24.

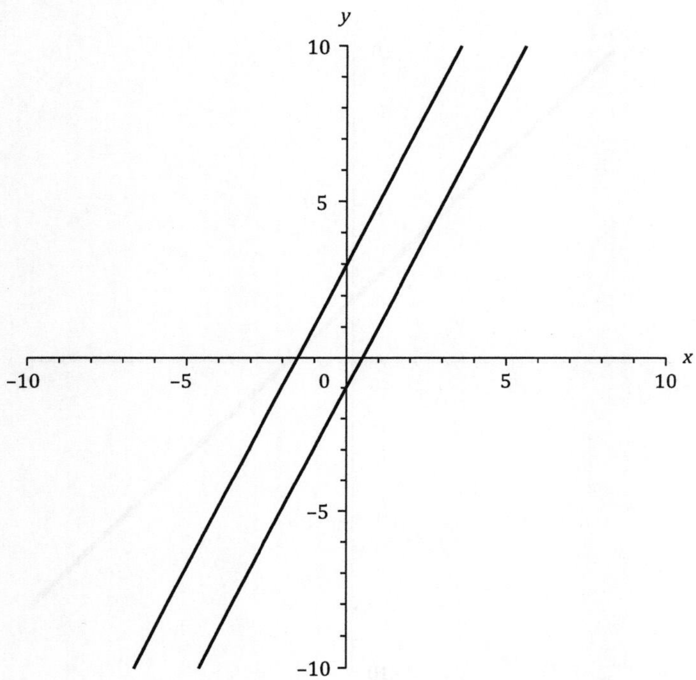

Fig. 3-24

In Option 1, we are talking about parallel lines.

Option 2: The lines coincide. For example, the lines $x + y = 2$ and $2x + 2y = 4$ are really the same equation. See Figure 3-25.

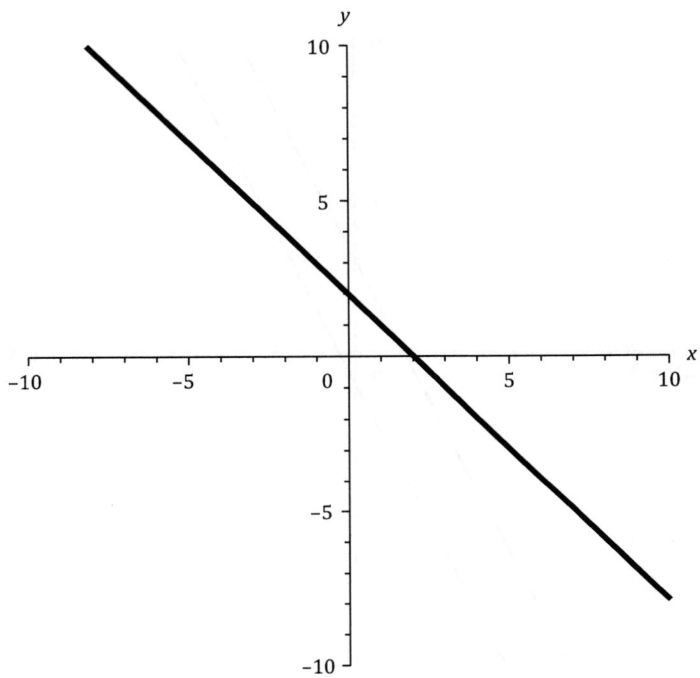

Fig. 3-25

Option 3: The lines intersect in one point. For example, note the lines $y = \frac{1}{2}x + 1$ and $y = -5x$. See Figure 3-26.

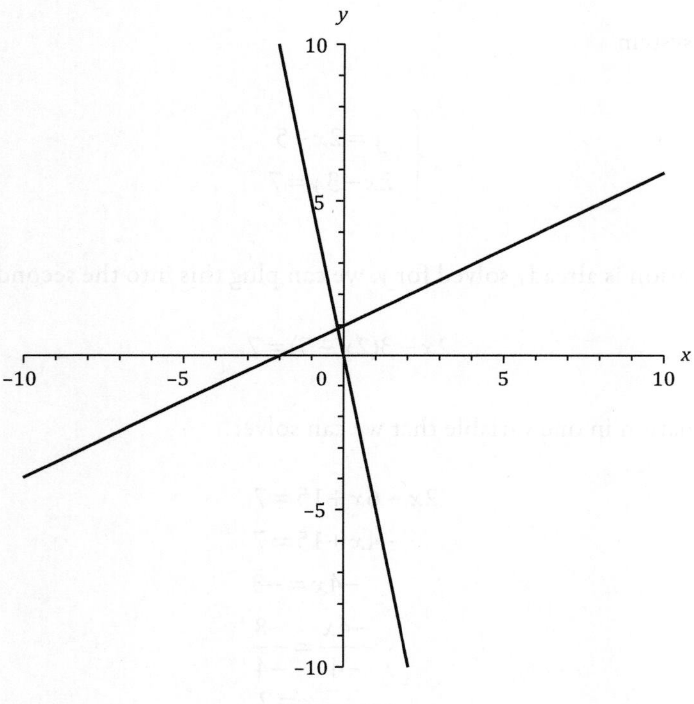

Fig. 3-26

What does this mean for the equations of the lines?

When finding where the graphs of two equations intersect, we solve them as a system of equations, meaning that we want to know what point (or points, if any) satisfies both equations for the same ordered pair. If we have two lines $Ax + By + C = 0$ and $Dx + Ey + F = 0$ and we set out to find their point of intersection, the notation for the system of equations to solve is

$$\begin{cases} Ax + By + C = 0 \\ Dx + Ey + F = 0 \end{cases}$$

We will discuss two methods for solving systems: substitution and elimination.

Substitution

If we solve one equation for a specific variable, plugging that into the other equation will get us an equation in one variable.

Ex) Consider the system

$$\begin{cases} y = 2x - 5 \\ 2x - 3y = 7 \end{cases}$$

Since the first equation is already solved for y, we can plug this into the second equation:

$$2x - 3(2x - 5) = 7.$$

This is now an equation in one variable that we can solve:

$$\begin{aligned} 2x - 6x + 15 &= 7 \\ -4x + 15 &= 7 \\ -4x &= -8 \\ \frac{-4x}{-4} &= \frac{-8}{-4} \\ x &= 2. \end{aligned}$$

This means that $x = 2$ is the x-coordinate of the point where these lines intersect. To find the solution of the system, we need the ordered pair. So our last step is to find y by plugging x into any equation that has y in it. Let's plug it back into the very first equation, $y = 2x - 5$:

$$\begin{aligned} &y = 2x - 5 \\ &y = 2(2) - 5 = 4 - 5 = -1. \end{aligned}$$

When $x = 2, y = -1$. So the ordered pair solution is $(2, -1)$. We should check to see that the point $(2, -1)$ lies on both lines.

For the first equation, we have $-1 = 2(2) - 5$, which is true.

For the second equation, we have $2(2) - 3(-1) = 7$, which is also true.

Also note that if we graph both lines, we can identify this as an intersection point. Graphing is the least reliable method of finding the solution to a system of equations. It is certainly possible, and it's easy if the lines have nice integer slopes and easy y-intercepts. But things can get complicated quickly, especially if we don't use an accurate scale when plotting the lines.

See Figure 3-27 for the graphs of the lines and the point of intersection.

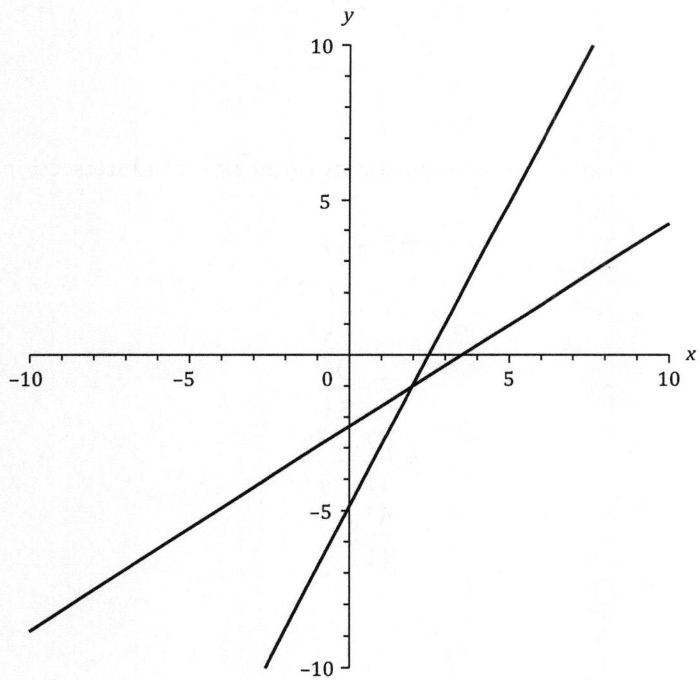

Fig. 3-27
Point of intersection (2, −1)

Example: Solve the system of equations

$$\begin{cases} x+4y=7 \\ 2x-3y=5 \end{cases}$$

Solution: In general, we should use substitution when we can easily solve for one of the variables. Notice that in the first equation in the system above, we can solve for x and have $x = 7 - 4y$. We can plug this into the second equation to obtain

$$\begin{aligned} 2(7-4y)-3y &= 5 \\ 14-8y-3y &= 5 \\ 14-11y &= 5 \\ -11y &= -9 \\ y &= \frac{9}{11}. \end{aligned}$$

Now, since $x = 7 - 4y$, we can find the x-coordinate of the point of intersection:

$$\begin{aligned} x &= 7-4y \\ x &= 7-4\left(\frac{9}{11}\right) \\ &= 7-\frac{36}{11} \\ &= \frac{77}{11}-\frac{36}{11} \\ &= \frac{41}{11}. \end{aligned}$$

Our solution is $\left(\frac{41}{11},\frac{9}{11}\right)$. It would be difficult to plot these two lines and identify the solution graphically. See Figure 3-28.

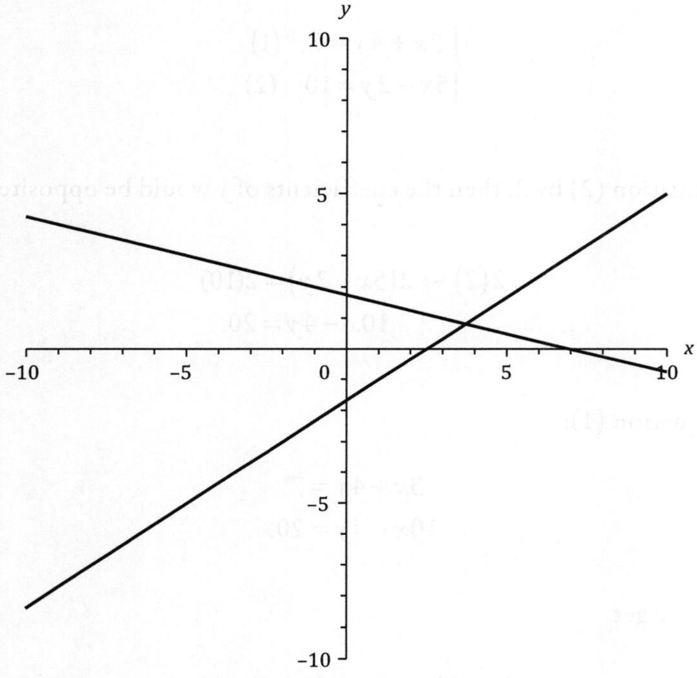

Fig. 3-28

Elimination

Elimination is used when one variable is not easily solved for. Luckily, with equations we can line up the coefficients so that one variable will cancel if we add the equations together. For example, if $a = b$ and $c = d$, then adding the equations gives us $a + c = b + d$.

If we were solving the system

$$\begin{cases} 3x+4y=7 \\ 5x-2y=10 \end{cases},$$

substitution would be a little more complicated. So using elimination, we pick one of the variables—let's pick y. Now, we want the coefficients to be *opposites*, so that if we add the equations together, y will cancel. Let's number the equations (1) and (2).

$$\begin{cases} 3x+4y=7 & (1) \\ 5x-2y=10 & (2) \end{cases}$$

If we multiplied equation (2) by 2, then the coefficients of y would be opposites:

$$\begin{aligned} 2(2) \rightarrow 2(5x-2y) &= 2(10) \\ 10x-4y &= 20. \end{aligned}$$

Let's add this to equation (1):

$$\begin{aligned} 3x+4y &= 7 \\ 10x-4y &= 20. \end{aligned}$$

Adding equations, we get

$$\begin{aligned} 3x+10x+4y-4y &= 7+20, \\ 13x &= 27. \end{aligned}$$

So $x=\frac{27}{13}$. Just plug this into an equation that has y in it—let's use equation (1):

$$\begin{aligned} 3\left(\frac{27}{13}\right)+4y &= 7 \\ \frac{81}{13}+4y &= 7 \\ 4y &= \frac{91}{13}-\frac{81}{13} \\ 4y &= \frac{10}{13} \\ y &= \frac{10}{13}\cdot\frac{1}{4} \\ y &= \frac{5}{26}. \end{aligned}$$

The solution is $\left(\frac{27}{13}, \frac{5}{26}\right)$.

YOU TRY IT: Check that $\left(\frac{27}{13}, \frac{5}{26}\right)$ is a solution of the system of equations

$$\begin{cases} 3x+4y=7 \\ 5x-2y=10 \end{cases}$$

Example: Solve this system of equations:

$$\begin{cases} 3x+7y=9 \\ -2x+5y=18 \end{cases}$$

Solution: In this system we can't just manipulate one equation. But note that the x's have opposites signs: in equation (1) x is positive, while in equation (2) x is negative. So let's pick x to eliminate. For the coefficients to match up, we will have to multiply the first equation by 2 and the second equation by 3:

$$\begin{aligned} 2\cdot(1) \rightarrow 2(3x+7y) &= 2\cdot 9 \\ 6x+14y &= 18. \end{aligned}$$

Multiplying equation (2) by 3 gives

$$\begin{aligned} 3\cdot(2) \rightarrow 3(-2x+5y) &= 3\cdot 18 \\ -6x+15y &= 54. \end{aligned}$$

Line up the equations:

$$\begin{aligned} 6x+14y &= 18 \\ -6x+15y &= 54. \end{aligned}$$

When we add these equations together, if one of the variables does not disappear, then we have done something wrong. But note that x will cancel if we add:

$$\begin{aligned} 6x - 6x + 14y + 15y &= 18 + 54 \\ 29y &= 72 \\ y &= \frac{72}{29}. \end{aligned}$$

Now that we have y, use any equation with x in it to find x—let's use equation (2):

$$\begin{aligned} -2x + 5\left(\frac{72}{29}\right) &= 18 \\ -2x + \frac{360}{29} &= 18 \\ -2x &= \frac{522}{29} - \frac{360}{29} \\ -2x &= \frac{162}{29} \\ x &= -\frac{1}{2}\left(\frac{162}{29}\right) \\ x &= -\frac{81}{29}. \end{aligned}$$

Our solution is $\left(-\frac{81}{29}, \frac{72}{29}\right)$.

For the last two cases, where the lines are parallel or coincide, what does this mean for their systems of equations? If two lines are parallel, they have the same slope but different y-intercepts.

Ex) Consider this system of equations:

$$\begin{cases} y = 2x - 1 \\ y = 2x + 5 \end{cases}$$

If we multiply equation (2) by -1 so that we can cancel, we have

$$-y = -2x - 5.$$

Add this to

$$y = 2x - 1,$$

and we're left with

$$\begin{aligned} -y + y &= -2x + 2x - 5 - 1 \\ 0 &= -6. \end{aligned}$$

What has happened here? We have come up with a contradiction, because $0 \neq -6$. This means the system of equations has no solution. So the lines will not intersect.

How about when the lines coincide?

Ex) Consider this system of equations:

$$\begin{cases} 2x + 10y = -7 \\ -4x - 20y = 14 \end{cases}$$

To use elimination, let's multiply the first equation by 2:

$$4x + 20y = -14.$$

Adding this to equation (2) yields

$$0 = 0$$

This is always true. No matter what ordered pair we pick that satisfies the first equation, it will automatically satisfy the second equation. So these lines coincide, and we have infinitely many solutions.

Section 3.4 Exercises

1. Solve this system of equations:

$$\begin{cases} x+y=1 \\ x-y=5 \end{cases}$$

2. Solve this system of equations:

$$\begin{cases} y=4x \\ 3x+10y=6 \end{cases}$$

3. Solve this system of equations:

$$\begin{cases} 3x-10y=5 \\ 2x=8 \end{cases}$$

4. Solve this system of equations:

$$\begin{cases} y=\frac{1}{2}x+1 \\ y=-5x \end{cases}$$

5. Solve this system of equations:

$$\begin{cases} x-5y=-6 \\ 3x+7y=26 \end{cases}$$

6. Solve this system of equations:

$$\begin{cases} 3x+4y=5 \\ -2x+7y=16 \end{cases}$$

7. Solve this system of equations:

$$\begin{cases} 2x+5y=-17 \\ 3x+10y=-38 \end{cases}$$

8. Solve this system of equations:

$$\begin{cases} 2x-3y=-5 \\ 4x+5y=12 \end{cases}$$

9. Solve this system of equations:

$$\begin{cases} 11x+2y=15 \\ -7x+4y=1 \end{cases}$$

10. Solve this system of equations:

$$\begin{cases} 3x-2y=10 \\ 2x+6y=3 \end{cases}$$

11. Solve this system of equations:

$$\begin{cases} x+2y=23 \\ 2x+3y=37 \end{cases}$$

12. Solve this system of equations:

$$\begin{cases} \frac{1}{2}x+y=5 \\ \frac{3}{2}x-2y=-15 \end{cases}$$

13. Solve this system of equations:

$$\begin{cases} 2x-3y=-18 \\ -3x+2y=7 \end{cases}$$

14. Determine if the following system of equations has no solutions, one solution, or infinitely many solutions:

$$\begin{cases} 3x-4y=8 \\ -6x+8y=-16 \end{cases}$$

15. Determine if the following system of equations has no solutions, one solution, or infinitely many solutions:

$$\begin{cases} -3x-3y=-21 \\ 2x+2y=22 \end{cases}$$

16. Determine if the following system of equations has no solutions, one solution, or infinitely many solutions:

$$\begin{cases} x+y=0 \\ x-y=0 \end{cases}$$

17. Determine if the following system of equations has no solutions, one solution, or infinitely many solutions:

$$\begin{cases} 2x+3y=8 \\ 2x+3y=7 \end{cases}$$

3.5: Three-Variable Systems of Linear Equations

Sometimes a problem can involve more than two variables, and often, one variable can depend on many things. For example, temperature is not just a function of time of day; it could also be a function of location, weather, any number of things. In this section we will solve three-variable systems. (We will also discuss applications of this in the next section.)

We will usually use x, y, and z as our three variables. Geometrically, the ordered triple (x, y, z) represents a point in three dimensions. This will be the solution to solving a system of three equations.

Most of the time, we will use elimination. However, we can use substitution if the coefficients of the variables are 1.

The process for this is similar to working with two equations (although we must be extra careful not to make any mistakes in the intermediate steps). With three variables, a systematic approach would be to eliminate one of the variables, like we did before. We'll follow these steps:

1. Number the equations, and pick two of them to eliminate one of the variables from.
2. Pick a different pair and eliminate the same variable.
3. This should create two new equations in only two variables. Proceed as in Section 3.4.

Which variable we start with does not matter, so search briefly for the one that looks easiest to cancel.

Example: Solve this system of equations:

$$\begin{cases} 2x+5y-6z=40 \\ x-y+2z=-9 \\ 3x+7y-z=34 \end{cases}$$

Solution: Let's number the equations:

$$\begin{cases} 2x+5y-6z=40 & (1) \\ x-y+2z=-9 & (2) \\ 3x+7y-z=34 & (3) \end{cases}$$

In the first two equations, any of the variables would not be difficult to eliminate. So let's choose to get rid of y because it has opposite signs. All we have to do is multiply equation (2) by 5:

$$\begin{aligned} 5\cdot(2) \rightarrow\ & 5(x-y+2z)=5(-9) \\ & 5x-5y+10z=-45. \end{aligned}$$

Add this to equation (1):

$$2x+5y-6z=40.$$

From this we obtain

$$\begin{aligned} 5x + 2x - 5y + 5y + 10z - 6z &= -45 + 40 \\ 7x + 4z &= -5 \quad (4). \end{aligned}$$

We are calling this equation (4). Label this clearly in your work, because we need to step away from this for the moment so we can eliminate y from a *different* pair of equations—either (1) and (3), or (2) and (3). Let's eliminate y from equations (2) and (3), because it looks much easier.

To continue, multiply equation (2) by 7:

$$\begin{aligned} 7 \cdot (2) \rightarrow 7(x - y + 2z) &= 7(-9) \\ 7x - 7y + 14z &= -63. \end{aligned}$$

Equation (3) is

$$3x + 7y - z = 34.$$

Adding these together, we have

$$\begin{aligned} 7x + 3x - 7y + 7y + 14z - z &= -63 + 34 \\ 10x + 13z &= -29 \quad (5). \end{aligned}$$

Once we have this new equation, make sure to label it (5). What we have is two equations in only x and z:

$$\begin{cases} 7x + 4z = -5 \quad (4) \\ 10x + 13z = -29 \quad (5). \end{cases}$$

There are not a lot of options here, so let's eliminate x. Multiplying equation (4) by -10, we have

$$\begin{aligned} -10 \cdot (4) \rightarrow -10(7x + 4z) &= -10(-5) \\ -70x - 40z &= 50. \end{aligned}$$

Now we have to multiply equation (5) by 7 to obtain opposite coefficients for x:

$$\begin{aligned} 7\cdot(5) \quad \rightarrow \quad 7(10x+13z) &= 7(-29) \\ 70x+91z &= -203. \end{aligned}$$

We are adding together

$$\begin{array}{r} -70x-40z=50 \\ \underline{70x+91z=-203,} \end{array}$$

with this result:

$$51z=-153.$$

Dividing both sides by 51, we have

$$z=-3.$$

Excellent! Now we need to get one of the other variables, so we go back to our work and find an equation with z and only one other variable. Let's use either (4) or (5). Using equation (4), we have

$$\begin{aligned} 7x+4(-3) &= -5 \\ 7x-12 &= -5 \\ 7x &= 12-5 \\ 7x &= 7 \\ x &= 1. \end{aligned}$$

Now we need to go all the way back to equation (1), (2), or (3) to use an equation that has y in it. Using equation (2), and using the fact that $x = 1$ and $z = -3$, we have

$$\begin{aligned} x - y + 2z &= -9 \\ (1) - y + 2(-3) &= -9 \\ 1 - y - 6 &= -9 \\ -y - 5 &= -9 \\ -y &= 5 - 9 \\ -y &= -4 \\ y &= 4. \end{aligned}$$

The solution to the system of equations is the ordered triple (1, 4, −3).

We mentioned before that an ordered triple determines a point in three dimensions. The general linear equation in *three* variables is

$$Ax + By + Cz + D = 0,$$

which graphs as a plane. For example, in Figure 3-29, the computer algebra software system Maple was used to graph $2x + 3y + z = 6$.

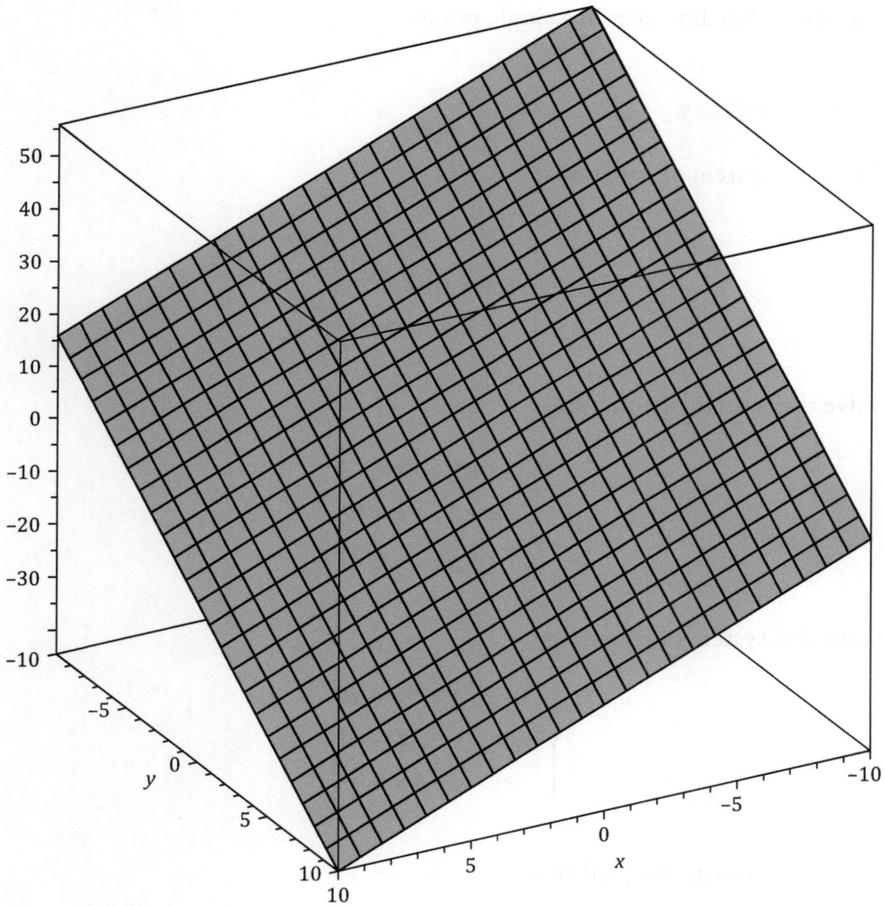

Fig. 3-29

Planes can be parallel, just like lines, so it's possible that a system could have no solution.

We have just seen an example of a system in where three planes intersect at precisely one point.

There can also be planes that coincide, to give infinitely many solutions. Three planes can also intersect in a line, again giving infinitely many solutions.

For now, we will concern ourselves with the algebra of finding one solution.

Later, we can talk about how three-variable systems graph as planes in three dimensions. We can use Maple to talk about the graphs of planes, and how there are examples where the planes can coincide, have no solution, intersect in a point, but also intersect in a line. Here, we will deal only with the case where they intersect in a single point.

Section 3.5 Exercises

1. Solve this system of equations:

$$\begin{cases} x+y-z=7 \\ x-y+z=5 \\ -x+y+z=-3 \end{cases}$$

2. Solve this system of equations:

$$\begin{cases} 2x+5y-z=-3 \\ 3x-2y+z=12 \\ 7x+y-2z=21 \end{cases}$$

3. Solve this system of equations:

$$\begin{cases} 5x+7y+9z=20 \\ -x+2y-3z=-11 \\ x+4y=-4 \end{cases}$$

4. Solve this system of equations:

$$\begin{cases} -3x+2y+4z=30 \\ 2x-5y+7z=-13 \\ -5x-6y+8z=-2 \end{cases}$$

5. Solve this system of equations:

$$\begin{cases} 2x+3y-5z=-29 \\ 3x-2y+6z=63 \\ 4x-7y+10z=120 \end{cases}$$

6. Solve this system of equations:

$$\begin{cases} x+2y=34 \\ 3y-4z=1 \\ 2x+5z=-16 \end{cases}$$

7. Looking ahead: Consider the equation of the plane $2x + 5y + 7z = 70$. We can expand the concept of intercepts from Section 3.1 to three dimensions. The ordered triple $(\alpha, 0, 0)$ is the x-intercept, because both y and z are zero. Similar definitions can be given for the y- and z-intercepts. Find the x-, y-, and z-intercepts of $2x + 5y + 7z = 70$.

3.6: Applications of Systems

In this section, we will discuss some of the many applications of systems of equations.

Let's talk first about mixtures, and how we were already doing a system of equations without knowing it.

Ex) Consider the problem we solved in Section 2.4: Suppose we need 25 liters of a 20% acid solution. Unfortunately, we have a liquid that is 10% acid and a liquid that is 30% acid. But, we can mix them together to create the required solution. How much of the liquids should we mix together?

If x is the amount of the 10% solution, and y is the amount of 30% solution, then we know that $x+y=25$ liters. We were already using substitution, because we immediately said the amount of the 30% solution must be $25-x$. We substituted this into the equation

$$0.1x + 0.3y = 0.2(25),$$

for y to obtain the equation that goes on to solve our problem.

Do a quick review of Section 2.4, where the same method of problem-solving was used. First, we want to know what the problem is asking for. Next, we'll define useful variables. Then, we will set up a system of equations that represent the information given in the problem, and solve.

Example: Suppose KC is rowing a rowboat gently down the stream. Downstream the current helps KC get to the destination in only 2 hours. However, rowing the same 10 miles upstream and fighting the current takes KC 6 hours. Assuming rates are constant, what is the speed of the current and how fast is KC rowing the boat?

Solution: To talk about distance, rate, and time, we use the $d = rt$ equation. The problem is asking for KC's rate of rowing and the speed of the current. Let's use those as the variables:

$$\begin{aligned} k &= \text{rate KC can row the boat,} \\ c &= \text{speed of the current.} \end{aligned}$$

We are talking about two situations: going upstream and going downstream. The distance for both trips is 10 miles. Going downstream takes 2 hours, and going upstream takes 6 hours. So the only question is the rate.

The rate of the boat going downstream will be the rate at which KC rows *plus* the speed of the current, because the current is helping. If distance equals rate times time, we have our first equation:

$$10 = (k + c)2.$$

Now, going upstream, we have the same distance of 10 miles, and the time is 6 hours. But the speed of the boat should be how fast KC is rowing *minus* the speed of the current, because now the current is working against KC. The equation for upstream is

$$10 = (k - c)6.$$

Here is our system of equations:

$$\begin{cases} 10 = (k + c)2 \\ 10 = (k - c)6 \end{cases}$$

Let's divide the first equation by 2, and the second equation by 6:

$$\begin{cases} k+c=5 \\ k-c=\dfrac{5}{3} \end{cases}$$

Now we can just add the equations together to find that

$$2k=5+\frac{5}{3},$$

or

$$2k=\frac{18}{3},$$

so $2k = 6$, and $k = 3$. To find c, let's use the first equation $k + c = 5$. We have $3 + c = 5$, so $c = 2$. KC is rowing at 3 mph, and the speed of the current is 2 mph.

Example: Leah took an exam and answered word problems and computation problems. Word problems take 8 minutes each and are worth 7 points if answered correctly. Computation problems take 5 minutes each and are worth 4 points if answered correctly. Leah answered everything correctly and scored 82 points in 98 minutes. How many of each type of problem did she answer?

Solution: If we let c be the number of computation problems answered, and let w be the number of word problems answered, then we are looking for c and w. The information we have is about the number of points scored and the number of minutes taken on the exam. So we have two equations: a "points" equation and a "time" equation. Since word problems are worth 7 points and computation problems are worth 4 points, and we know Leah scored 82 points, we must have that

$$7w + 4c = 82.$$

Regarding time, word problems take 8 minutes and computation problems take 5 minutes. Since Leah took 98 minutes on the exam, we have

$$8w + 5c = 98.$$

Thus our system of equations is this:

$$\begin{cases} 7w + 4c = 82 & (1) \\ 8w + 5c = 98 & (2) \end{cases}$$

Let's choose to eliminate c. Multiplying equation (1) by -5 gives

$$\begin{aligned} -5\cdot(1) \quad \rightarrow \quad & -5(7w+4c) = -5(82) \\ & -35w - 20c = -410. \end{aligned}$$

To match up the coefficient of c, we multiply equation (2) by 4:

$$\begin{aligned} 4\cdot(2) \quad \rightarrow \quad & 4(8w+5c) = 4(98) \\ & 32w + 20c = 392. \end{aligned}$$

We are adding

$$\begin{array}{r} -35w - 20c = -410 \\ \underline{32w + 20c = 392} \end{array}$$

and we obtain

$$-3w = -18,$$

so $w = 6$. Using this in equation (1), we have

$$7(6) + 4c = 82,$$

which means that

$$4c = 82 - 42.$$

Finally, $4c = 40$, or $c = 10$. Leah answered 10 computation problems and 6 word problems.

Example: Computer chip A can perform 8 operations per second, computer chip B can perform 15 operations per second, and computer chip C can perform 10 operations per second. Chip A costs \$4 to make, chip B costs \$10 to make, and chip C costs \$7 to make. The number of A chips used was exactly the same as the sum of the number of B and C chips used. If \$158 was spent making chips and 259 operations were performed, how many of each chip were used?

Solution: This will be a system of three equations. Let a equal the number of A chips used, let b equal the number of B chips used, and let c equal the number of *C* chips used.

The number of operations performed was 259, so we must have that $8a + 15b + 10c = 259$. The cost of the chips is $4a + 10b + 7c$, which was \$158. And finally, we know that the number a must be the same as the sum of b and c, so $a = b + c$. The following is our system of equations:

$$\begin{cases} 8a + 15b + 10c = 259 & (1) \\ 4a + 10b + 7c = 158 & (2) \\ a = b + c & (3) \end{cases}$$

To eliminate a in the first two equations, let's substitute the third equation in both of them. Plugging equation (3) into equation (1) gives us equation (4):

$$\begin{aligned} 8(b + c) + 15b + 10c &= 259 \\ 8b + 8c + 15b + 10c &= 259 \\ 23b + 18c &= 259 \quad (4). \end{aligned}$$

To get another equation, with only b and c, plug equation (3) into equation (2):

$$\begin{aligned} 4(b + c) + 10b + 7c &= 158 \\ 4b + 4c + 10b + 7c &= 158 \\ 14b + 11c &= 158 \quad (5) \end{aligned}$$

So our system with only b and c is this:

$$\begin{cases} 23b + 18c = 259 & (4) \\ 14b + 11c = 158 & (5) \end{cases}$$

Now, let's cancel c. Multiply equation (4) by -11:

$$-11 \cdot (4) \rightarrow \quad -11(23b + 18c) = -11(259)$$
$$-253b - 198c = -2849$$

Now multiply equation (5) by 18:

$$18 \cdot (5) \quad \rightarrow \quad 18(14b + 11c) = 18(158)$$
$$252b + 198c = 2{,}844$$

We are adding

$$\begin{array}{r} -253b - 198c = -2849 \\ \underline{252b + 198c = 2844} \\ \\ -b = -5. \end{array}$$

So $b = 5$. Now equation (4)—or equation (5); we just need an equation that has two variables in it—to solve for c:

$$\begin{aligned} 23(5) + 18c &= 259 \\ 115 + 18c &= 259 \\ 18c &= 144 \\ c &= 8. \end{aligned}$$

Now, use equation (3) to find a:

$$a = 5 + 8 = 13.$$

Thus, 13 A chips were used, 5 B chips were used, and 8 C chips were used.

Section 3.6 Exercises

1. James is traveling 3,000 miles from Los Angeles to New York on a plane. The trip to New York takes 6 hours, because there is a wind at the back of the plane. The trip back to Los Angeles, traveling against the headwind, takes 10 hours. What is the speed of the wind, and what is the speed of the airplane with no wind?
2. The music department sold 125 tickets to the school musical. Student tickets cost $1 and parent tickets cost $2. If the department made $200 in profit, how many students went to the musical and how many parents went?
3. Pine mulch is available in bags that cover 2 cubic feet of a garden, and cypress mulch is available in bags that cover 3 cubic feet. Pine bags cost $5, and cypress bags cost $6. Elizabeth likes to mix the two in her garden. She spent $238 and was able to cover 106 cubic feet of her garden. How many bags of each kind of mulch did Elizabeth buy?
4. Eddie is paddling a canoe upstream against the current for 4.5 hours, and it takes him 4.5 hours. Paddling downstream with the current takes Eddie 2 hours. If he travels 6 miles on each trip, how fast is he paddling and what is the speed of the current?
5. Emma is traveling on a boat from Baltimore to New York. On the way there, a current helps the trip take only 5 hours, but on the way back, against the current, the trip takes 9 hours. If the journey from Baltimore to New York is 200 miles by boat, what is the speed of the boat (with no current) and what is the speed of the current?
6. Diana takes Vitamin B and Vitamin C pills. Each Vitamin B pill weighs 2 grams and each vitamin C pill weighs 4 grams. Each packet of Vitamin B contains 2 pills and each packet of Vitamin C contains 8 pills. If Diana takes 52 grams of pills and uses 9 packets, how many of each kind of pill did Diana take?
7. Local land-line phone calls used to cost $0.50 per minute, and long-distance land-line phone calls cost $2 per minute. If a phone bill showed 65 minutes of call time and a charge of $62.50, how many of those minutes were local and how many were long-distance?
8. For a particular flight, an airline sold first-class tickets for $1,000 each and coach tickets for $300 each. If the airline seated 80 people and made $34,500, how many of each type of ticket was sold?
9. The company's yearly budget for printers and paper is $240. Last year, the company bought eight times as many reams of paper as they bought printers. If printers cost $40 each and reams of paper cost $2.50 each, how many of each were bought?
10. Bellamy's math team bus is on its way from Fort Myers to Orlando. The trip to Orlando takes 4 hours. On the way back to Fort Myers, the driver George decides to go 10 mph faster. If the trip home takes 3.5 hours, then how fast did George drive on the way to Orlando, and how far is it from Fort Myers to Orlando?

11. On Brianna's flight to New York, there is a tail wind helping the plane get the 1,000-mile journey in 6 hours. On Brianna's return flight, the headwind makes the journey 8 hours long. Assuming the speed of the plane is constant, what is the speed of the plane in still air and what is the speed of the wind?
12. Daniela bought frogs, mice, and octopi for her students to dissect in biology class. She paid the medical supply company $1 per frog, $2 per mouse, and $4 per octopus, for a total of $110. She bought twice as many mice as frogs, and she bought ¾ of the amount of mice in octopi. How many of each did she get for her students?
13. At a local concert venue, children's tickets cost $2, teenagers spend $5 per ticket, and adults spend $8 per ticket. If 600 people attended the most recent concert, $4,050 was spent on tickets, and for every 1 child there were 8 adults, how many children attended? How many teenagers? How many adults attended?

Chapter 3 Review

3-A

1. Give the x- and y-intercepts of the line $-2x + 9y = 30$, and graph the line.
2. Find the slope of the line through $(-5, -10)$ and $(6, -2)$.
3. Write the equation of the line through $(-8, 5)$ with slope 2 in slope-intercept form, and graph the line.
4. Write the equation of the line parallel to $x + y = 7$ that passes through $(2, 16)$ in slope-intercept form.
5. Solve this system of equations:

$$\begin{cases} x = 3y + 5 \\ 3x - 4y = 6 \end{cases}$$

6. Solve this system of equations:

$$\begin{cases} 2x + 5y = 10 \\ 3x - 4y = -6 \end{cases}$$

7. Markers cost 10 cents to produce, and paperclips cost 2 cents to produce. One machine that produces either markers or paperclips made 50 objects in a day. If this cost 308 cents (or $3.08), then how many of each object were made?

3-B

1. Dan rows upstream in half an hour against the current, and he rows downstream in one-quarter of an hour with the current. If he travels 2 miles each way, what is the speed of Dan's boat in still water, and what is the speed of the current?
2. Solve this system of equations:

$$\begin{cases} 3x-y+2z=-3 \\ 2x+y+5z=-15 \\ x+5y-3z=19 \end{cases}$$

3. Line l has equation $14x + 6y = 30$, and line m has equation $3x - 17y = 12$. Line n has the same y-intercept as line l and line n has the same x-intercept as line m. Write the equation of line n in slope-intercept form.
4. Write the equation of the line perpendicular to $3x - 4y = -15$ that passes through (18, −15) in slope-intercept form.
5. Solve this system of equations:

$$\begin{cases} \frac{1}{2}x+\frac{3}{4}y=10 \\ \frac{1}{3}x-\frac{1}{5}y=15 \end{cases}$$

6. Write the equation of the line through (−1, −6) and (−5, −30) in slope-intercept form.
7. Solve this system of equations:

$$\begin{cases} 2x-y=4 \\ 3x+5y=10 \end{cases}$$

CHAPTER 4

Inequalities

In this chapter we study inequalities, which opens up our mathematical dictionary even wider. A quick example: If we know that we need at least 14 minutes to complete a task, then our time t must satisfy $t \geq 14$. If we know that we have at most 40 minutes to complete the task, then we also have $t \leq 14$.

The inequalities > and < represent "greater than" and "less than," respectively.

The inequalities ≥ and ≤ represent "greater than or equal to" and "less than or equal to," respectively.

Sometimes the inequalities < and > are called **strict inequalities** because they do not allow the possibly of equality (whereas ≥ and ≤ allow this possibility).

4.1: Simple Inequalities

When we are looking at an inequality, we are still asking "For what x is this valid?" The answer can be different from what we saw with equations. Instead of *usually* finding *one* solution to an equation, inequalities *usually* have *infinitely many* solutions. Of course, there are exceptions.

Ex) Consider the inequality $x > 2$. To solve this inequality, x just has to be any real number larger than 2. So $x = 3$ works, $x = \pi$ works, etc. This is why when we solve inequalities, the solution is a shaded number line. See Figure 4-1. Note the open circle at 2. We shade to the right of 2 to indicate that any number greater than 2 is a solution.

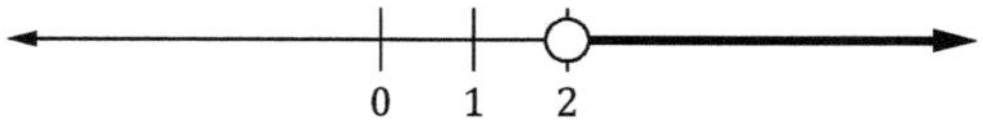

Fig. 4-1

Ex) Likewise, the inequality $x < -5$ has a solution that can be shown graphically. We shade a number line to the left of −5, but with an open circle at −5. See Figure 4-2.

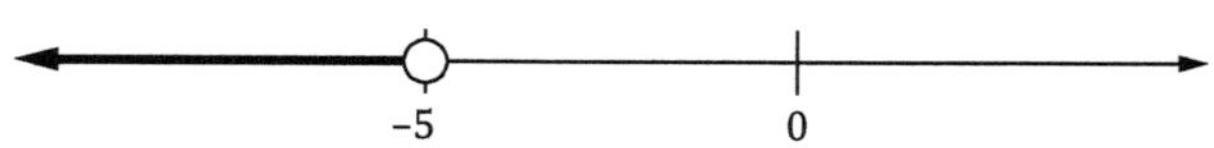

Fig. 4-2

Ex) Now, if the inequality we are looking at is $x \geq 0$, then we close the circle at 0 and shade to the right. See Figure 4-3.

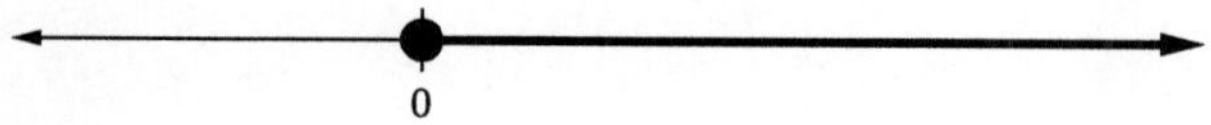

Fig. 4-3

Ex) And finally, if we have $x \leq 10$, we shade to the left of 10 but we make sure the circle is closed at 10. See Figure 4-4.

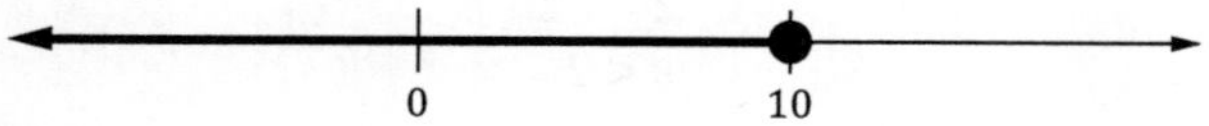

Fig. 4-4

The way we write down inequalities is not commutative. If $x < a$, then it is not true that $a < x$.

But saying $x < a$ is equivalent to saying that $a > x$. We just have to make sure that if we flip the inequality sign, we also flip what we write on each side. We did not have to worry about this with equations, so we have to use more care.

Ex) $x > 3$ is not the same thing as $3 > x$.

Ex) Saying $x \geq -\frac{1}{4}$ is equivalent to saying $-\frac{1}{4} \leq x$.

The rules for inequalities are mostly identical to those for equations. Addition and subtraction can be done freely (as long as we do the same to both sides), but the rules change with multiplication and division of negative numbers. If we are solving an inequality that requires multiplication or division by a negative number, we must flip the sign.

Ex) If we ask, "What x solves the inequality $-x > 4$?," then we are asking for values of x that have opposites larger than 4. So the solution could be -5, -10, etc. But we cannot divide both sides of the inequality $-x > 4$ by -1 and keep the sign the way it is; this would be saying that $x > -4$, which is false if the solutions are -5, -10, and smaller numbers. So we must flip the sign to get $x < -4$.

Example: Solve the inequality $2x + 1 < 5$ and graph the solution.

Solution: Proceeding as we did when solving an equation, we begin by subtracting 1 from both sides:

$$\begin{gathered} 2x+1<5 \\ -1 \quad -1 \\ 2x<4 \\ \frac{2x}{2}<\frac{4}{2} \\ x<2. \end{gathered}$$

For a graph, see Figure 4-5.

Fig. 4-5

Example: Solve $-3x + 5 \geq -2(x - 10)$ and graph the solution.

Solution: We proceed as normal, except when we multiply both sides by -1:

$$\begin{gathered} -3x+5 \geq -2x-2(-10) \\ -3x+5 \geq -2x+20 \\ +2x \quad +2x \\ -x+5 \geq 20 \\ -5 \quad -5 \\ -x \geq 15 \end{gathered}$$

It is at this step that we have to be careful, and flip the inequality sign:

$$\begin{gathered} -1(-x) \leq -1(15) \\ x \leq -15. \end{gathered}$$

See Figure 4-6.

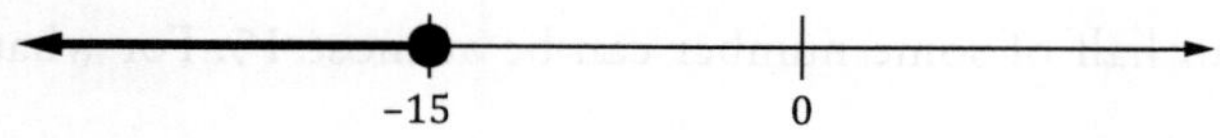

Fig. 4-6

We must also be careful that inequalities make sense. Just like equations, there are inequalities with no solution and inequalities with infinitely many solutions.

Example: Solve the following inequality, and graph the solution on a number line:

$$2(2x - 5) - 14 \leq 4(x - 6).$$

Solution: We begin by simplifying:

$$\begin{gathered} 4x - 10 - 14 \leq 4x - 24 \\ 4x - 24 \leq 4x - 24 \\ +24 \quad +24 \\ 4x \leq 4x \\ -4x \quad -4x \\ 0 \leq 0 \end{gathered}$$

Note that in the second step of the solution, we have the same thing on both sides of the inequality. Whenever this happens, the inequality will be *always* true or *never* true. And since zero is less than or equal to zero, this inequality is always true. So the solution is the entire real number line. See Figure 4-7.

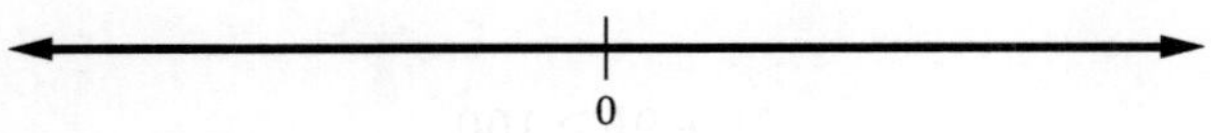

Fig. 4-7

Example: Solve the inequality $2x > 2x + 1$.

Solution: If we subtract $2x$ from both sides, we end up with $0 > 1$, a statement that is never true. Remember, we are asking "For what x this inequality will hold?" And the answer is that for no x is it true. If we graphed this solution, we would draw a number line with nothing circled or shaded.

Applications

Example: 8 more than half of some number can be at most 15. For what numbers could this statement be?

Solution: The key here is to recognize that "at most" means "less than or equal than." With "8 more than" we would be adding 8 to our number, which is some variable x. So we have

$$8+\frac{1}{2}x\leq 15.$$

Solving this inequality gives

$$\begin{aligned}\frac{1}{2}x&\leq 7\\ 2\left(\frac{1}{2}x\right)&\leq 2(7)\\ x&\leq 14.\end{aligned}$$

So the number can be anything less than or equal to 14.

Example: Jake makes \$9 an hour at work lifting boxes, and he gets a \$25 stipend each shift he works. If Jake needs to make at least \$100 in one shift, how many hours does he have to work?

Solution: "At least" means "greater than or equal to," because Jake can make \$100 or more (but not less). Now, he makes a fixed \$25 each shift, plus he makes \$9 for every hour. If we let h be the number of hours worked, then we have

$$25+9h\geq 100.$$

Subtracting 25 from both sides gives

$$9h\geq 75.$$

And if we divide both sides by 9, we have

$$h\geq\frac{75}{9}$$

or

$$h \geq \frac{25}{3}.$$

Now, since $\frac{25}{3} = 8.\overline{3}$, Jake must work at least $8\frac{1}{3}$ hours. If he can work only a whole number of hours, he must work 9 hours (since 8 hours will not get us to $100).

Section 4.1 Exercises

1. Graph the following inequality on a number line: $x \geq 0$.
2. Graph the following inequality on a number line: $x < -\frac{3}{2}$.
3. Graph the following inequality on a number line: $x > 8$.
4. Graph the following inequality on a number line: $x \leq 2$.
5. Solve the following inequality and graph the solution on a number line: $x + 5 < 4$.
6. Solve the following inequality and graph the solution on a number line: $x - 5 \geq 1$.
7. Solve the following inequality and graph the solution on a number line: $3x - 2 > 7$.
8. Solve the following inequality and graph the solution on a number line: $\frac{1}{2}x - 5 \leq 2$.
9. Solve the following inequality and graph the solution on a number line:

$$3x + 5 < 2x - 1.$$

10. Solve the following inequality and graph the solution on a number line:

$$2x - 5 \geq 3(x - 5).$$

11. Solve the following inequality and graph the solution on a number line:

$$\frac{1}{2}(4x - 5) > 6x.$$

12. Solve the following inequality and graph the solution on a number line:

$$3(5 - x) - 1 \leq 2 + 4(1 - 2x).$$

13. Solve the following inequality and graph the solution on a number line:

$$\frac{1}{2}-\frac{3}{2}(6x-1)\geq\frac{1}{2}(1-2x)-5.$$

14. Solve the following inequality and graph the solution on a number line:

$$0.4(x-10)\leq 1+0.3(9x+20).$$

15. Solve the following inequality and graph the solution on a number line:

$$3(x-6)<6\left(\frac{1}{2}x-3\right).$$

16. Solve the following inequality and graph the solution on a number line:

$$3(x-6)\leq 6\left(\frac{1}{2}x-3\right).$$

17. Solve the following inequality and graph the solution on a number line:

$$2(x-4)+1>2x-8.$$

18. Solve the following inequality and graph the solution on a number line:

$$2\left(\frac{1}{2}x-5\right)\leq 4\left(\frac{1}{4}x-\frac{5}{2}\right)-1.$$

19. A certain hotel charges \$8 to access the Internet and \$15 for every gigabyte (GB) of data used. If a guest sets aside \$70 for Internet use, how many whole GB can be downloaded to guarantee spending at most \$70?
20. It costs \$1 to access the computer lab, and \$0.75 for every hour of use of a computer. At most, how many hours will Amanda be able to use a computer in the lab if she can spend at most \$4?
21. A company charges \$60 to rent a space for a party, and then charges a certain amount per hour that the space is rented. If the company knows a certain party is planning to

use the space for 5 hours, how much should the company charge per hour to make more than \$400?

22. It takes Josh 2 hours to do the reading assignment. Then, each essay response question takes him 30 minutes to complete. How many whole questions will Josh be able to answer after reading the assignment if he has only $4\frac{3}{4}$ hours in which to do homework?

4.2: Compound Inequalities

When we're discussing inequalities, the words "and" and "or" are very important. In general, for x to be one thing "or" another is less restrictive than if we required x to satisfy one condition "and" another. When either of these words is used with inequalities, we call them <u>compound inequalities</u>.

Ex) For instance, in the inequalities $x < 4$ or $x > 10$, we're saying that x can be less than 4 or greater than 10. Graphically, the solution to this problem has two branches. We circle 4 and 10 on a number line, then shade to the left of 4 and to the right of 10. See Figure 4-8.

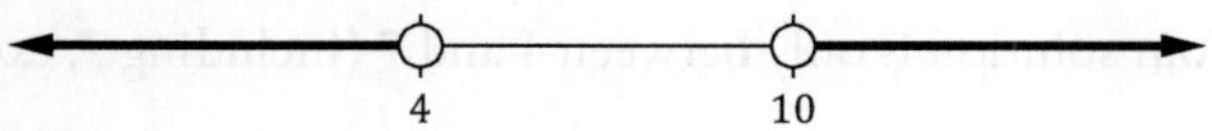

Fig. 4-8

Ex) Similarly, to graph the solution to $x \geq -1$ or $x < -2$, we draw a number line with a filled-in circle at -1 and an open circle at -2. Then we take the numbers greater than or equal to -1, and shade to the right of -1 and to the left of -2. See Figure 4-9.

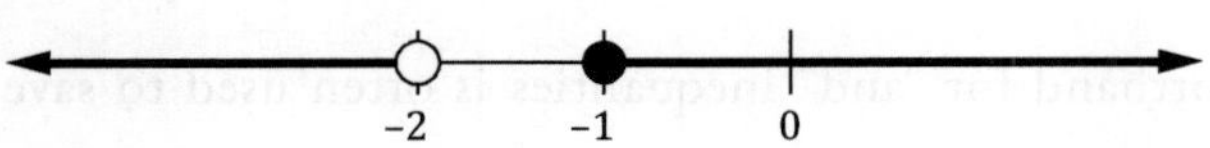

Fig. 4-9

In terms of sets, "and" means <u>intersection</u>. Literally, we will look for the places where the solutions of our two inequalities intersect (run over each other), and this will be the solution.

Ex) Consider the problem where $x > 4$ and $x \leq 7$. We have an open circle at 4 and a closed circle at 7. Now, consider this one inequality at a time. If $x > 4$, we shade to the right of 4. See Figure 4-10.

Fig. 4-10
x > 4

Then, if $x \leq 7$, we shade to the left of 7. When shading, we mark just above where we shaded before, so we can tell the difference. See Figure 4-11.

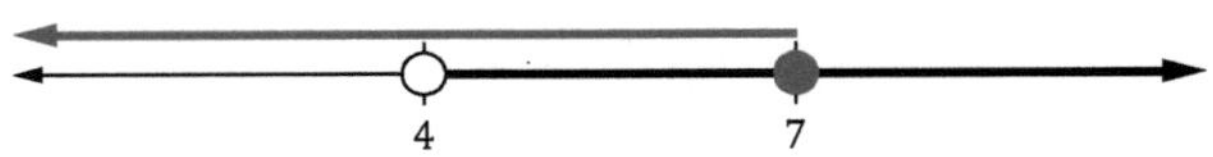

Fig. 4-11
x > 4 graphed with x <= 7

The actual solution to our inequalities will be the place where these are both true *at the same time*. In this example, our solution is only between 4 and 7 (including 7, excluding 4). See Figure 4-12.

Fig. 4-12
x > 4 and x <= 7

In mathematics, a shorthand for "and" inequalities is often used to save space. Consider this definition:

$$a < x < b \quad \text{is equivalent to} \quad a < x \ \text{ and } \ x < b.$$

For example, just above we were looking for the area where $x > 4$ and $x \leq 7$. If we write this so the inequality goes in the same directions, we have $4 < x$ and $x \leq 7$. When we can put the pieces together, we see that this is the same as $4 < x \leq 7$.

Example: Solve the inequality and graph the solution: $-1 \leq 4x + 5 < 7$.

Solution: Let's break the inequality into

$$-1 \le 4x + 5 \qquad \text{and} \qquad 4x + 5 < 7.$$

Whenever we split up the work like this, it is important to keep track. Sometimes it helps to work one inequality and then the other.

$$\begin{aligned} -1 \le 4x+5 \quad &\text{and} \quad 4x+5<7. \\ -5 \quad -5 \quad & \qquad -5 \quad -5 \\ -6 \le 4x \quad &\text{and} \quad 4x < 2 \\ \frac{1}{4}(-6) \le \frac{1}{4}(4x) \quad &\text{and} \quad \frac{1}{4}(4x) < \frac{1}{4}(2) \\ -\frac{3}{2} \le x \quad &\text{and} \quad x < \frac{1}{2}. \end{aligned}$$

If we put this back into the form $-\frac{3}{2} \le x < \frac{1}{2}$, we see that x must be between those two numbers, with an open circle at $\frac{1}{2}$ and a closed circle at $-\frac{3}{2}$. See Figure 4-13.

Fig. 4-13

Unfortunately, there is no shorthand for an "or" inequality. It is vital that we do not try to combine "or" inequalities. That can be done only with an "and" inequality.

Example: Solve $3x - 8 > 7$ or $\frac{1}{2}x + 6 < \frac{11}{2}$.

Solution: Proceeding carefully, we have

$$\begin{aligned}
3x-8>7 \quad &or \quad \frac{1}{2}x+6<\frac{11}{2}\\
+8 \quad +8 \quad & \quad\quad -6 \quad -6\\
3x>15 \quad &or \quad \frac{1}{2}x<\frac{11}{2}-\frac{12}{2}\\
3x>15 \quad &or \quad \frac{1}{2}x<-\frac{1}{2}\\
\frac{3x}{3}>\frac{15}{3} \quad &or \quad 2\left(\frac{1}{2}x\right)<2\left(-\frac{1}{2}\right)\\
x>5 \quad &or \quad x<-1.
\end{aligned}$$

See Figure 4-14 for the graph of the solution.

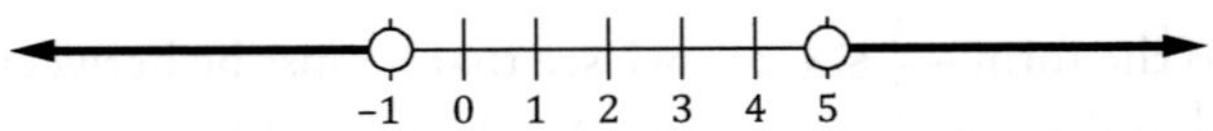

Fig. 4-14

Example: Solve the inequality $3x + 10 < 13$ or $-x - 7 \leq -5$ and graph the solution.

Solution: We have

$$\begin{aligned}
3x+10<13 \quad &or \quad -x-7\leq -5\\
-10 \quad & \quad\quad +7\\
3x<3 \quad &or \quad -x\leq 2\\
\frac{3x}{3}<\frac{3}{3} \quad &or \quad -1(-x)\geq -1(2)\\
x<1 \quad &or \quad x\geq -2.
\end{aligned}$$

See Figure 4-15 for the graph of the solution. If x can be any number less than 1 or any number greater than or equal to -2, then x can be *any* real number. In terms of sets, "or" means union. We "union" the pieces together to get the solution.

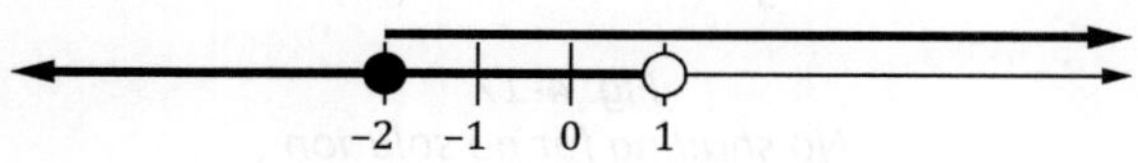

Fig. 4-15
All real numbers work

Example: Solve the inequality $2x < 10$ and $5x - 1 \geq 29$.

Solution: Solving, we have

$$2x < 10 \quad \textit{and} \quad 5x - 1 \geq 29$$
$$\frac{2x}{2} < \frac{10}{2} \quad \textit{and} \quad 5x - 1 \geq 29$$
$$x < 5 \qquad\qquad +1 \quad +1$$
$$x < 5 \quad \textit{and} \quad 5x \geq 30$$
$$x < 5 \quad \textit{and} \quad \frac{5x}{5} \geq \frac{30}{5}$$
$$x < 5 \quad \textit{and} \quad x \geq 6.$$

Now here is where we have to be careful. We have an open circle at 5 and a closed circle at 6. But when we shade for an "and" inequality, the solution will be where they *overlap*. Note that there are no numbers less than 5 and also greater than or equal to 6. See Figure 4-16.

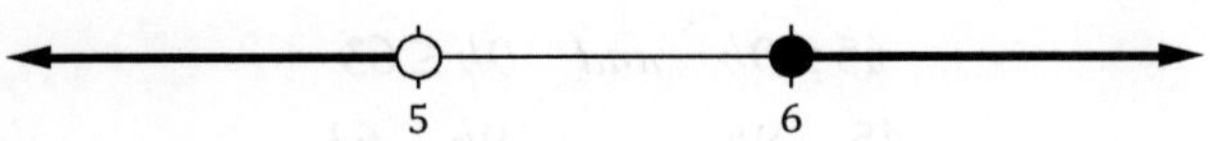

Fig. 4-16
Solution is overlap, so there is no solution

So there is no solution to this inequality, and we can represent this with an unshaded number line. See Figure 4-17.

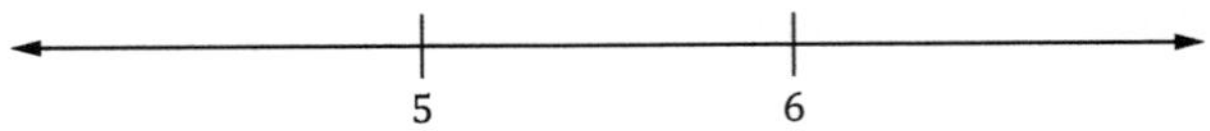

Fig. 4-17
No shading for no solution

Example: Jake makes a \$25 stipend as well as \$9 an hour during every shift he works. If Jake needs to make at least \$70 in one shift, but the company he works for will not let him make more than \$88 in one shift, what is the range for the number of hours he should work in one shift to satisfy both requirement?

Solution: The amount of money Jake makes in the shift can be at most \$88, but it must be at least \$70. Put in terms of inequalities, we have

$$70 \leq \text{amount made in one shift} \leq 88.$$

Let's use h as the number of hours worked. Then each shift Jake makes a fixed \$25 as well as $9h$ dollars, for a total of $9h + 25$ dollars. So we have

$$70 \leq 9h + 25 \leq 88.$$

Splitting the inequality apart, we have

$$\begin{array}{rcl} 70 \leq 9h + 25 & \textit{and} & 9h + 25 \leq 88 \\ -25 \quad -25 & & -25 \quad -25 \\ 45 \leq 9h & \textit{and} & 9h \leq 63 \\ \frac{45}{9} \leq \frac{9h}{9} & \textit{and} & \frac{9h}{9} \leq \frac{63}{9} \\ 5 \leq h & \textit{and} & h \leq 7. \end{array}$$

So Jake must work at least 5 hours, but at most 7 hours. Any number of hours worked between 5 and 7 will satisfy all requirements.

Section 4.2 Exercises

1. Graph the solution to $x < 2$ or $x \geq 5$.
2. Graph the solution to $x < 2$ and $x > -1$.
3. Graph the solution to $x \geq 3$ and $x \leq 6$.
4. Graph the solution to $x \leq 1$ or $x \geq 0$.
5. Solve $7 < 2x - 1 < 11$ and graph the solution.
6. Solve $16 \geq 3x + 4 > 10$ and graph the solution.
7. Solve $-13 \leq \frac{1}{2}(x - 4) < 5$ and graph the solution.
8. Solve $15 > \frac{1}{6}(12 - x) \geq -1$ and graph the solution.
9. Solve $2x < 4$ or $5x - 1 > 49$ and graph the solution.
10. Solve $7x + 2 > 51$ or $6 - 5x > 101$ and graph the solution.
11. Solve $\frac{1}{2}(5 - x) < -10$ or $3 - \frac{1}{3}x > -2$ and graph the solution.
12. Solve $0.1x - 3 \leq -2.5$ or $0.1x - 5 \geq -4$ and graph the solution.
13. Solve $7x + \frac{1}{2} < -\frac{27}{2}$ and $9 \leq 3(x + 2)$ and graph the solution.
14. Solve $3x - 5 < -2$ or $15(x - 1) + 2 > -13$ and graph the solution.
15. Graph the solution to $x < 7$ or $x < 10$.
16. Graph the solution to $x < 7$ and $x < 10$.
17. Jessica's dog Spoon catches 6 lizards a day. If Jessica says she has seen at least 24 different lizards caught but no more than 30, then what is the range on the number of days that Spoon has been catching lizards?
18. Kyle had driven 30 different cars before 2005, and for every year after 2005 he drove 8 more different cars per year. He says that he has driven at least 54 different cars, but at most 110. What is the range on the number of years Kyle has been driving cars since 2005?
19. When solving "and" inequalities, we do not have to split them apart. Consider $3 < 2x - 1 \leq 11$. If we add 1 to all three parts of the inequality, we have $4 < 2x \leq 12$. Next, we divide all terms by 4. We have

$$\frac{4}{2} < \frac{2x}{2} \leq \frac{12}{2},$$

which solves to $2 < x \leq 6$. Use this strategy to solve $3 \leq 4x - 5 < 15$.

20. Use the strategy in the previous problem to solve $17 > 7 - 5x > 12$.

4.3: Absolute Value Inequalities

Recall from when we studied absolute value equations that there were two things to do. The same is true with absolute value inequalities.

If $x < 3$ and $x > -3$, then we can write this as a string of inequalities, like

$$-3 < x < 3.$$

Since $|3| = 3$ and $|-3|$ is also equal to 3, this is our definition:

$$|x| < a \text{ is equivalent to } -a < x < a.$$

Consider $|x| < 3$. Translating this to words, recall that absolute value means "distance from zero." So if the distance between x and 0 is less than 3, then x could be any number less than 3 as long as it is also any number greater than -3. See Figure 4-18.

Fig. 4-18
Solution to |x|< 3

So this is how $|x| < 3$ means $-3 < x < 3$. Or we could write it as $x < 3$ and $x > -3$.

Now consider if the inequality were flipped. If we were asking for the x's that make the statement $|x| > 3$ true, the distance between x and the origin would have to be at least 3 (excluding 3). So we could pick anything larger than 3, *or* we could pick anything less than -3. See Figure 4-19 for the graph of the solution to $|x| > 3$.

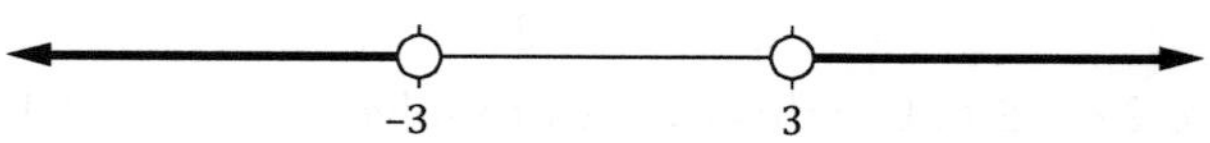

Fig. 4-19
Solution to |x|> 3

We solve absolute value inequalities as follows:

1. Write one inequality with the absolute value bars dropped.
2. Write another inequality that has the sign flipped, and take the opposite of one side of the inequality.
3. Solve as an "and" or "or" inequality, depending on which solution makes sense for the problem.

Example: Solve $\left|\frac{1}{2}(x-3)-5\right| \leq 4$.

Solution: We think of this as an "and" inequality, because the distance between what is inside the absolute value bars and zero can be at most 4. So we have

$$\frac{1}{2}(x-3)-5 \leq 4 \quad \textit{and} \quad \frac{1}{2}(x-3)-5 \geq -4$$

$$+5 \quad +5 \qquad\qquad +5 \quad +5$$

$$\frac{1}{2}(x-3) \leq 9 \quad \textit{and} \quad \frac{1}{2}(x-3) \geq 1$$

$$2\cdot\frac{1}{2}(x-3) \leq 2\cdot 9 \quad \textit{and} \quad 2\cdot\frac{1}{2}(x-3) \geq 2\cdot 1$$

$$x-3 \leq 18 \quad \textit{and} \quad x-3 \geq 2$$

$$+3 \quad +3 \qquad +3 \quad +3$$

$$x \leq 21 \quad \textit{and} \quad x \geq 5.$$

The solution of this inequality is the numbers between 5 and 21, inclusive. See Figure 4-20.

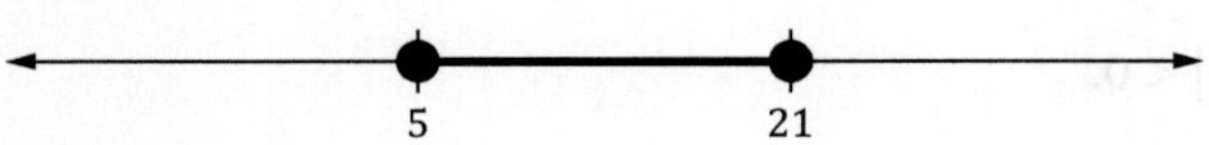

Fig. 4-20

Example: Solve $|1-3(2-x)|>8$.

Solution: We write one inequality with the absolute value bars dropped, and another inequality where we do the flip/opposite. Now, since the inequality is a "greater than" symbol, what's inside the absolute value bars can be greater than 8 or less than −8. So this is an "or" inequality. We have

$$\begin{aligned} 1-3(2-x)&>8 \quad \textit{or} \quad 1-3(2-x)<-8 \\ 1-3(2)-3(-x)&>8 \quad \textit{or} \quad 1-3(2)-3(-x)<-8 \\ 1-6+3x&>8 \quad \textit{or} \quad 1-6+3x<-8 \\ -5+3x&>8 \quad \textit{or} \quad -5+3x<-8 \\ +5 \quad +5& \qquad\qquad +5 \quad +5 \\ 3x&>13 \quad \textit{or} \quad 3x<-3 \\ x&>\frac{13}{4} \quad \textit{or} \quad x<-1. \end{aligned}$$

Since $\frac{13}{4}=3+\frac{1}{4}=3.25$, on our number line we shade above 3.25 or below −1, with open circles. See Figure 4-21.

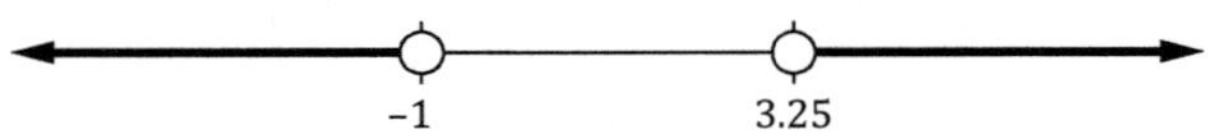

Fig. 4-21

When we solve absolute value inequalities, it is important to know whether it is an "and" or an "or." Consider the following two examples.

Example: Solve $|x+1|<0$.

Solution: This is why keeping the "and" is important here. We have $x+1<0$ and $x+1>0$.

$$\begin{aligned} x+1<0 \quad &\textit{and} \quad x+1>0 \\ x<-1 \quad &\textit{and} \quad x>-1. \end{aligned}$$

No single number can be both less than -1 *and* greater than -1. So there is no solution to this inequality.

Note the difference in the following example.

Example: Solve $|x + 1| \geq 0$.

Solution: Using the same method, we write one inequality with the absolute value bars dropped, and then we flip the sign and take the opposite of the right-hand side (which will still be 0). We have

$$x+1 \geq 0 \quad or \quad x+1 \leq 0$$
$$x \geq -1 \quad or \quad x \leq -1.$$

We are allowing x to be greater than or equal to -1, *or* less than or equal to -1. But now we are describing any real number, so the solution here is the entire real number line.

It's not always obvious, but in the previous two examples we can step back and look at what is being asked for. The problem of finding when $|x+1| < 0$ is the problem of finding when an absolute value of something (a distance) can be negative. The answer is *never*.

On the other hand, if we are asking for when an absolute value is nonnegative (for example, if we're solving $|x+1|$ ‡ 0), then the answer is *always*.

Section 4.3 Exercises

1. Solve $|x| < 5$ and graph the solution.
2. Solve $|x| > 5$ and graph the solution.
3. Solve $|x| < -5$ and graph the solution.
4. Solve $|x| > -5$ and graph the solution.
5. Solve $|2x| < 4$ and graph the solution.
6. Solve $|2x-1| \leq 5$ and graph the solution.
7. Solve $|1-x| < 4$ and graph the solution.
8. Solve $|3x-5| \leq 20$ and graph the solution.
9. Solve $\left|\frac{1}{2}x-5\right| < 16$ and graph the solution.
10. Solve $|4+2(x-1)| \leq 10$ and graph the solution.
11. Solve $4+2|x-1| < 18$ and graph the solution.
12. Solve $|7x-10| > 4$ and graph the solution.
13. Solve $|2+x| \geq 5$ and graph the solution.
14. Solve $|x-1| > 10$ and graph the solution.
15. Solve $\left|4\left(x-\frac{1}{2}\right)\right| \geq 15$ and graph the solution.
16. Solve $\left|\frac{1}{2}x+8\right| \geq 0$ and graph the solution.
17. Solve $\left|\frac{1}{3}x-2\right| \geq 1$ and graph the solution.
18. Solve $|2x+1| \geq 8$ and graph the solution.
19. Solve $|3x-5| \geq 10$ and graph the solution.
20. Solve $4+|1-2x| > 1$ and graph the solution.
21. Solve $4-|1-x| > 1$ and graph the solution.
22. Think about it: Explain why $|x-2| \geq 0$ has infinitely many solutions, but $|x-2| \leq 0$ only has one solution.

4.4: Linear Inequalities

In this section, we will be solving inequalities with two variables. For example, when we are solving the inequality

$$y > x + 1,$$

we are asking what ordered pairs (x, y) satisfy the inequality $y > x + 1$. For example, the ordered pair $(2, 4)$ works because if $x = 2$ and $y = 4$ then it is true that $4 > 2 + 1$. The ordered pair $(-6, 0)$ also works, because if $x = -6$ and $y = 0$, we have $0 > -6 + 1$.

We can probably guess a few more, but there are infinitely many ordered pairs that work. To find them, we apply the following procedure.

Here are the steps for solving linear inequalities:

1. Replace the inequality with an equals sign. Graph the corresponding line as dashed if the inequality is strict ($<$ or $>$); otherwise, graph the corresponding line as solid ($\leq$ or $\geq$).
2. This cuts the Cartesian plane in half. To test the inequality, pick a point not on the line. If this ordered pair works, shade the half-plane that this point lies in. If the ordered pair does not work, shade the other half-plane.

Example: Graph the solution to $y > x + 1$.

Solution: The corresponding line is $y = x + 1$, which has a slope of 1 and y-intercept 1. Since the inequality is >, the first step is to graph $y = x + 1$ as a dashed line. See Figure 4-22.

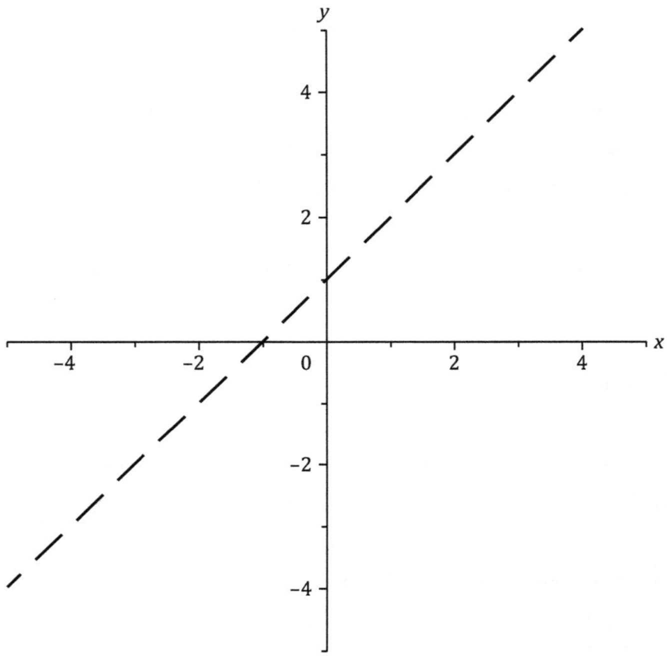

Fig. 4-22

Next, we test a point that's not on the line. We should pick an ordered pair so that the computation is easy, and (0, 0) is a good choice. If $x = 0$ and $y = 0$, we want to know if the original inequality $y > x + 1$ is true. Plugging in the values, we see that

$$0 > ?0 + 1$$
$$0 > ?1,$$

which is not true. So we shade on the side of the line that the origin is not on. See Figure 4-23.

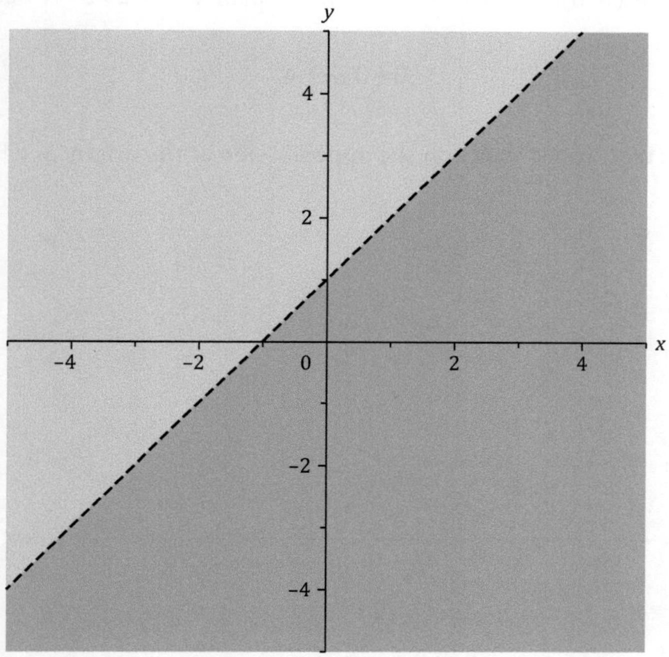

Fig. 4-23

If we want to know what ordered pairs (x, y) solve the inequality $y > x + 1$, we look at the region above the line $y = x + 1$. We can pick any ordered pair in that shaded region, and it will solve the inequality. So the solution to a single linear inequality will be what's called a <u>half-plane</u>.

Example: Graph the solution to $3x + 2y \leq -4$.

Solution: We replace the inequality with an equals sign, so we need to graph $3x + 2y = 4$. Putting this in slope-intercept form, we solve for y:

$$\begin{aligned} 2y &= -3x - 4 \\ y &= -\frac{3}{2}x - 2. \end{aligned}$$

The line has a slope of $-\frac{3}{2}$ and a y-intercept of -2. Since the original inequality was "≤," we draw the line as a solid line. Next, we test a point in the original inequality. Since the origin does not lie on the line, let's test (0, 0). When $x = y = 0$, the inequality $3x + 2y \leq -4$ becomes

$$0 + 0 \leq ? -4.$$

Note that this is not true, so we shade on the opposite side of the origin. See Figure 4-24.

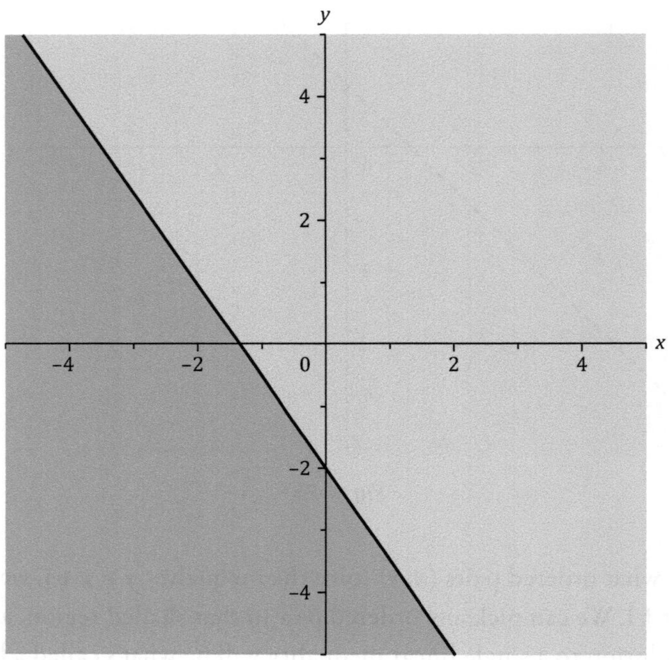

Fig. 4-24

A nice trick is to look at the inequality in slope-intercept form to determine which way to shade. If we have

$$y \geq mx + b,$$

then we shade above the line—because we want the y-values to be greater than $mx + b$. If we have

$$y < mx + b,$$

then we shade *below* the line. We want the ordered pairs where y is less than $mx + b$.

Vertical and Horizontal Lines

Example: Graph the solution to $y < 4$.

Solution: The line $y = 4$ is a horizontal line with zero slope hitting the y-axis at $(0, 4)$. Now, using the trick from above, since we want the y's to be less than 4 in our solutions, we shade below the line. Alternatively, we can still plug in the origin to this inequality. If $x = y = 0$, the inequality $y < 4$ becomes $0 <? 4$, which is true. So we shade on the same side of the line as the origin, which is again below the dashed graph of $y = 4$. See Figure 4-25.

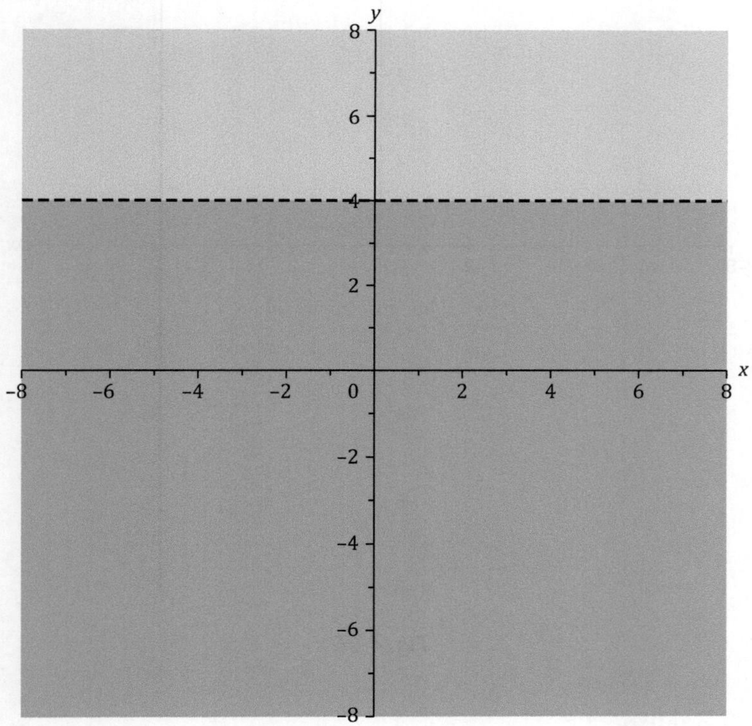

Fig. 4-25

Example: Graph the solution to $x \geq 5$.

Solution: The line $x = 5$ is a vertical line passing through $(5, 0)$. If we test the origin to see which side of the line to shade, we will obtain that it's the opposite side. But also note that we are looking for ordered pairs where the x-coordinate is at least 5. So this should be to the right of the line. See Figure 4-26.

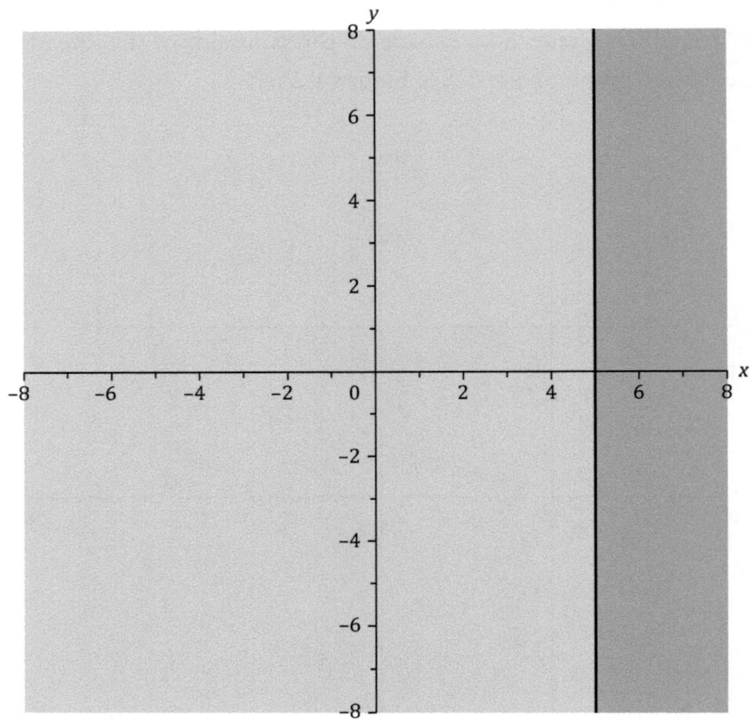

Fig. 4-26

Section 4.4 Exercises

Graph the solution to the following inequalities.

1. $y < 2x - 5$
2. $y \geq 2x - 5$
3. $y > -\frac{1}{2}x + 1$
4. $y \leq -3x - 2$

5. $2x+3y>5$
6. $x-2y<0$
7. $x+2y<1$
8. $3x-5y\geq 15$
9. $-3x-2y\leq 1$
10. $3x+4y-5\leq 0$
11. $2x-y>0$
12. $2x+2>0$
13. $y\leq -1$
14. $x>\frac{1}{2}$
15. $3y+2\geq 0$
16. $5x+12y-24<0$
17. $4x+7y>0$
18. $x<2$
19. $y\geq 0.5$
20. $y\leq \frac{1}{4}x-1$

4.5: Systems of Inequalities

The process for solving systems of inequalities is more like the work we did in the previous section (Section 4.4) rather than in Chapter 3 when we were solving systems of equations. That is, it can be helpful to solve the corresponding system of equations, but it may not be necessary every time.

Our strategy in this section will be to graph each inequality like we did in Section 4.4. If it is a strict inequality, we draw a dashed line; if it's not, we draw a solid line. Then we use the methods from Section 4.4 to determine which side of the line to shade. Now, the solution to the *system* of inequalities will be any area where we shade twice. This is otherwise known as the intersection of the graphs. So wherever the graph is double-shaded, we can pull any ordered pair to satisfy both inequalities at the same time.

Example: Graph the solution to this system of inequalities:

$$\begin{cases} y \geq 2x - 3 \\ y < -\dfrac{1}{2}x + 1 \end{cases}$$

Solution: First, we graph $y = 2x - 3$ as a solid line. It has slope 2 and y-intercept 3. The graph is shown in Figure 4-27.

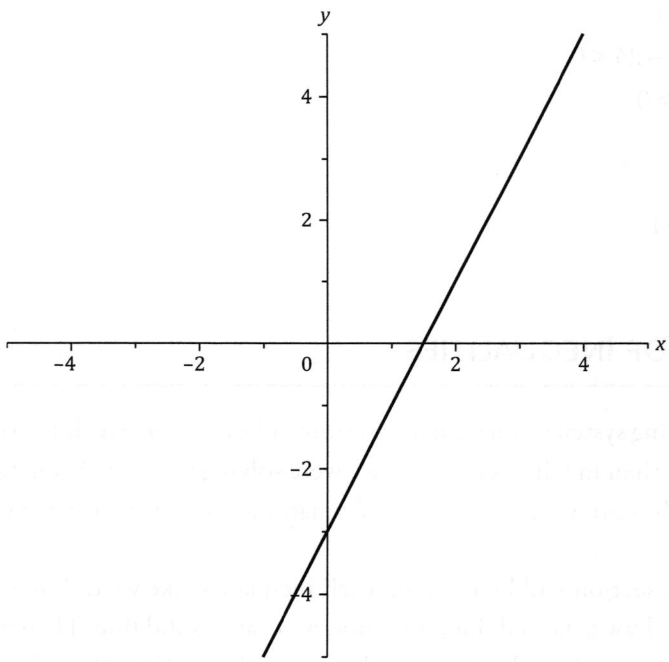

Fig. 4-27

Next, we shade above the line because it is a ≥ inequality. Alternatively, note that the origin will solve the inequality, so we shade on the same side as the origin. See Figure 4-28.

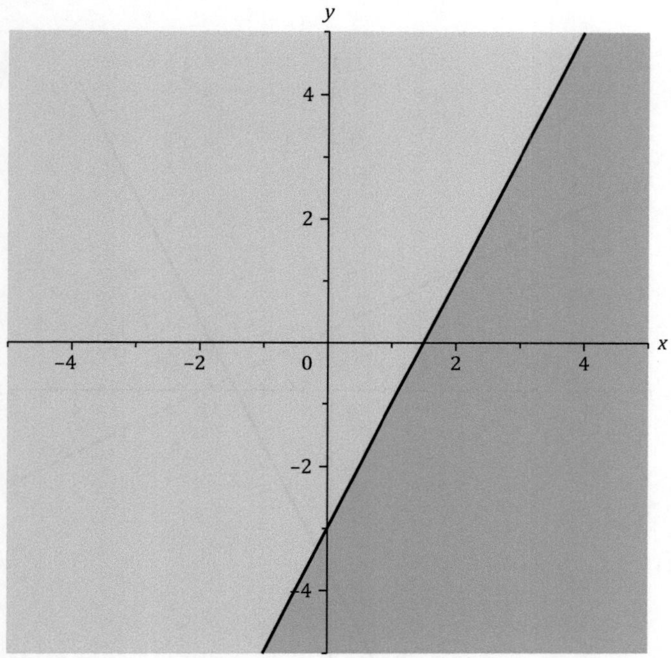

Fig. 4-28

Then we graph $y = -\frac{1}{2}x + 1$ as a dashed line with slope $-\frac{1}{2}$ and y-intercept 1. Since this is a < inequality, we shade below this particular line. The final solution will be wherever we have shaded twice. The solution is graphed in Figure 4-29.

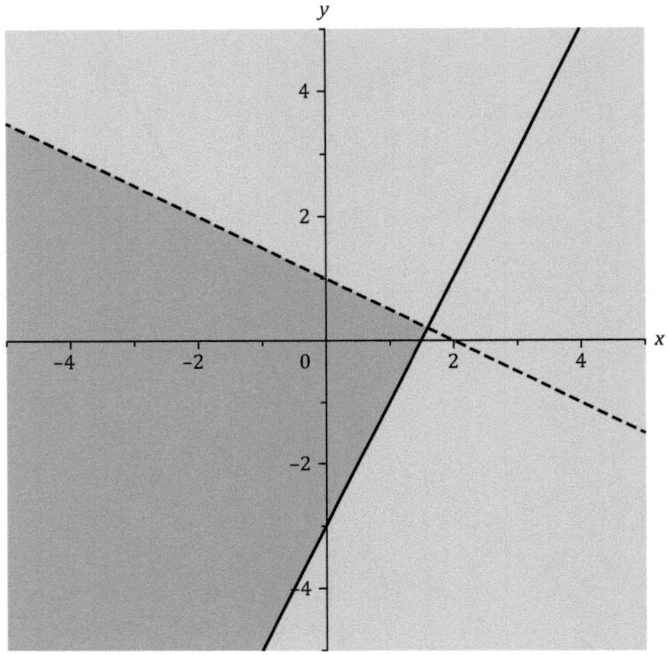

Fig. 4-29

Example: Graph the solution to this system of inequalities:

$$\begin{cases} y < 2 \\ x \geq -3 \end{cases}$$

Solution: First, we graph the line $y = 2$ as a dashed line. This is a horizontal line passing through (0, 2). Since we need y-values less than 2, we will shade below this line. See Figure 4-30.

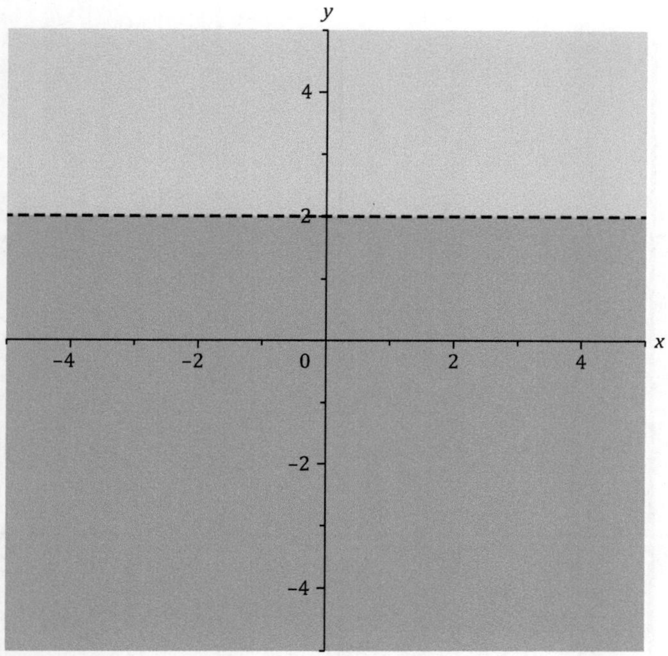

Fig. 4-30

Next, we draw the line $x = -3$, which is a vertical line through $(-3, 0)$. Since we want ordered pairs where $x \geq -3$, we shade to the right of the vertical line. The solution to this system is any place where we have shaded twice in this process. The graph is given in Figure 4-31.

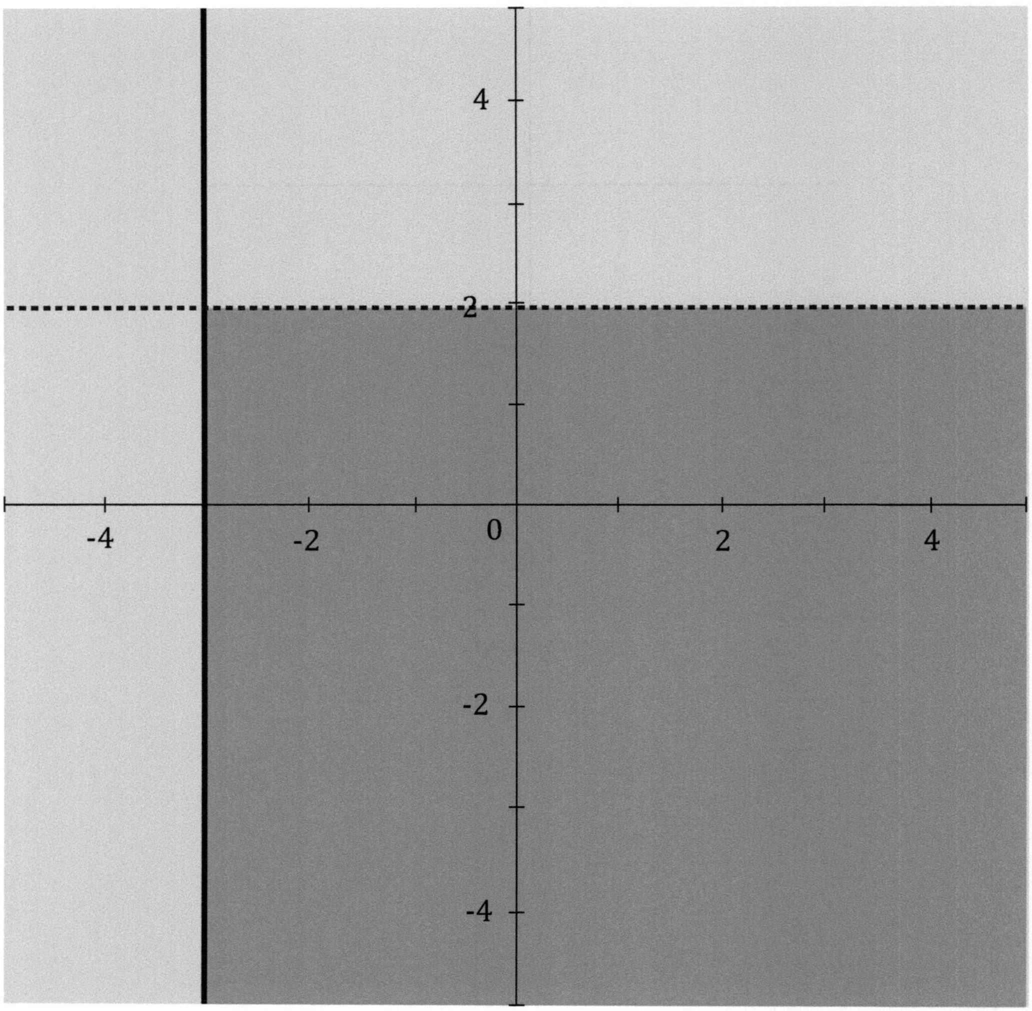

Fig. 4-31

Example: Graph the solution to this system of inequalities:

$$\begin{cases} y \geq \dfrac{1}{4}x + 1 \\ y < \dfrac{1}{4}x - 1 \end{cases}$$

Solution: First, let's graph the line $y = \frac{1}{4}x + 1$. It will be solid, with slope $\frac{1}{4}$ and y-intercept 1, and we'll shade above this line because of the ≥. See Figure 4-32.

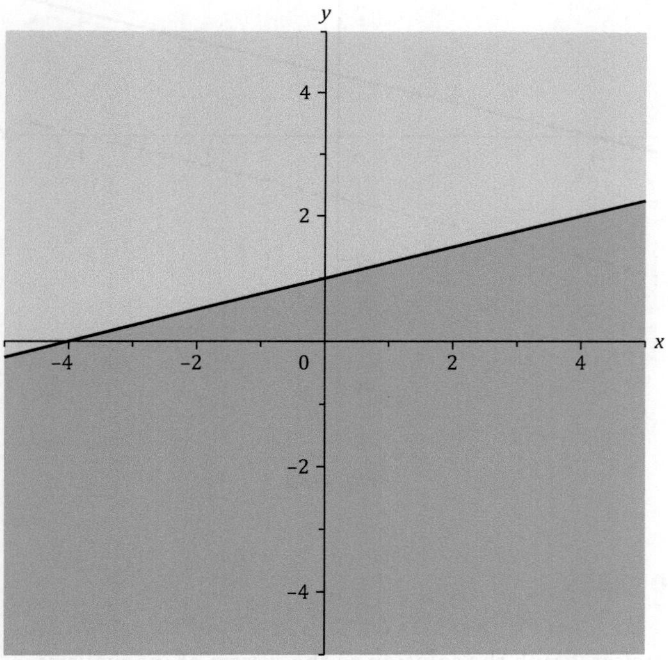

Fig. 4-32

Next, we graph $y = \frac{1}{4}x - 1$. It will be a dashed line with slope $\frac{1}{4}$ and y-intercept -1, and we will shade below the line because of the < direction of the inequality. However, note that we are not shading over any previous shading. These lines are parallel, and we are shading on the opposite sides. So there will be no solution to this inequality. See Figure 4-33.

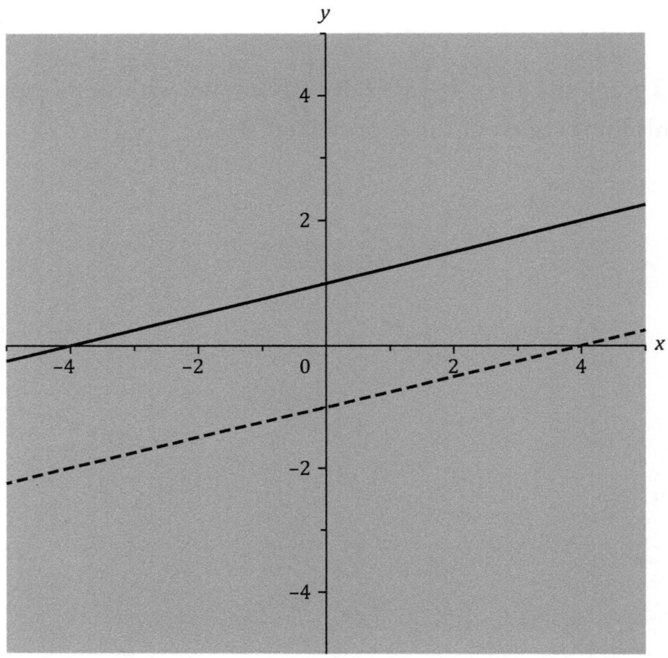

Fig. 4-33

Section 4.5 Exercises

In the following exercises, graph the solution to the system of inequalities.

1. $\begin{cases} y \geq x - 1 \\ y < -x - 1 \end{cases}$

2. $\begin{cases} y < x - 1 \\ y > -x - 1 \end{cases}$

3. $\begin{cases} y > x - 1 \\ y \geq -x - 1 \end{cases}$

4. $\begin{cases} y > \frac{1}{3}x \\ y \leq 2 - x \end{cases}$

5. $\begin{cases} 2x - y < 1 \\ \frac{1}{2}x - y \geq -1 \end{cases}$

6. $\begin{cases} -x + 4y < 12 \\ x + 2y \geq -8 \end{cases}$

7. $\begin{cases} 2x + 2y < 1 \\ x < 1 \end{cases}$

8. $\begin{cases} 11x - 33y \leq 3 \\ 2x + y > 0 \end{cases}$

9. $\begin{cases} y + 1 < 0 \\ x - 3 \geq 0 \end{cases}$

10. $\begin{cases} 3x + 6y < 7 \\ 2x - y \geq 1 \end{cases}$

11. $\begin{cases} 7x - 4y \geq 10 \\ 2x + 10y < 5 \end{cases}$

12. $\begin{cases} y > x \\ y < x - 1 \end{cases}$

13. $\begin{cases} y > x \\ y < -1 \end{cases}$

14. $\begin{cases} 3x > 5 \\ x + y \leq 2 \end{cases}$

15. $\begin{cases} x - 11y < 22 \\ 11y \leq x + 11 \end{cases}$

16. The graph in Figure 4-34 is the solution to a system of inequalities. Which of the following ordered pairs is a solution to this system?
    a. (–49,563)
    b. (618,–2)
    c. (–579,864)
    d. (2,–1,672)

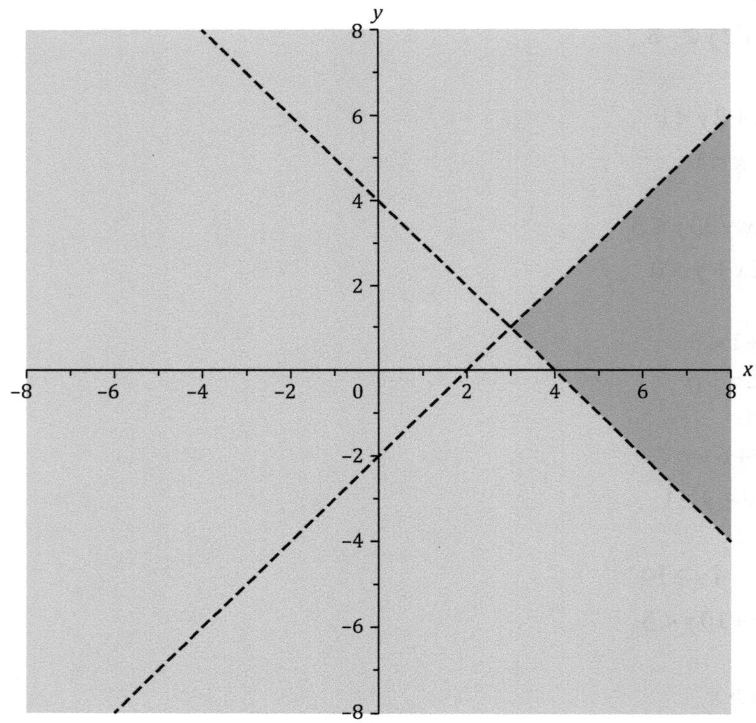

Fig. 4-34

17. The graph in Figure 4-35 is the solution to a system of inequalities. Which of the following ordered pairs is a solution to this system?
    a. (146,1,347)
    b. (−246,352)
    c. (4,−895)
    d. (−657,−10)

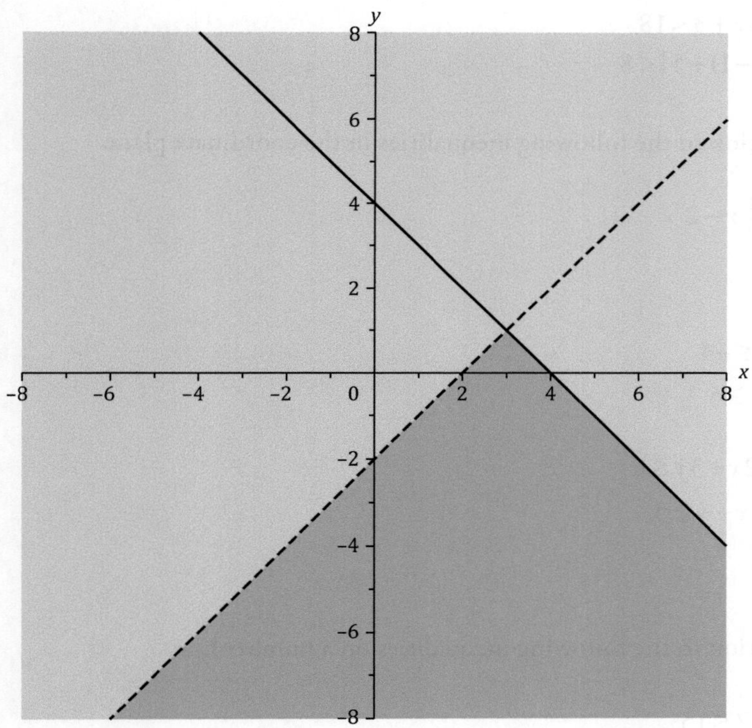

Fig. 4-35

CHAPTER 4 REVIEW

4-A

Graph the solution to the following inequalities on a number line.

1. $x > 2$
2. $3x - 5 > 7$ or $2x - 1 \leq -3$
3. $2 < 5x + 3 < 18$
4. $|3(x-1)+5| < 8$

Graph the solution to the following inequalities in the coordinate plane.

5. $y \geq \frac{1}{2}x - 2$
6. $x > 3$
7. $\begin{cases} y \leq -4 \\ x \geq 5 \end{cases}$
8. $\begin{cases} -2x + 3y \leq 6 \\ x + y \geq 5 \end{cases}$

4-B

Graph the solution to the following inequalities on a number line.

1. $3x - 7 \leq 15$
2. $-3x + 10 < -11$ or $-10x + 7 > -53$
3. $\frac{1}{2}x - 10 > -10$ or $-3x + 29 > 26$
4. $3(x+2) > 6$ and $-6x - 10 > -10$
5. $-62 \leq 10x - 2 \leq 28$
6. $\left|3\left(x - \frac{1}{2}\right) + 1\right| > 6$
7. $|3x - 1| < -1$

Graph the solution to the following inequalities in the coordinate plane.

8. $7x-5y\geq 3$
9. $y>0$
10. $\begin{cases} y>1 \\ x\leq -1 \end{cases}$
11. $\begin{cases} 4x+y\leq 5 \\ x+y<0 \end{cases}$
12. $\begin{cases} x+2y\geq 6 \\ 2x+4y<-4 \end{cases}$

CHAPTER 5

Polynomials

In this chapter, we will study polynomials, which are sums of natural-number powers of one variable. The definition of a polynomial in one variable looks like

$$y = a_n x^n + a_{n-1} x^{n-1} + \cdots + a_1 x + a_0$$

where $a_0, a_1, a_2, \ldots, a_{n-1}, a_n$ are all real numbers and $n \in \mathbb{N}$. We usually write the powers in descending order, but this is not necessary.

Of course, there are polynomials in many variables. Expressions like

$$x^2 y^3 + xy^4$$

and

$$3xyz + 3x^2 yz + 6xy^2 z + 9xyz^2$$

are polynomials as well.

5.1: Operations on Polynomials

In this section, we will discuss adding, subtracting, and multiplying polynomials.

Adding and Subtracting Polynomials

Adding and subtracting polynomials is using like terms.

Ex) If we add $(4x^2 + 2x - 1) + (3x^2 - x + 1)$, we group the x^2's together, the x's together, and the constants together. So we have

$$\begin{aligned}(4x^2 + 2x - 1) + (3x^2 - x + 1) &= 4x^2 + 3x^2 + 2x - x - 1 + 1 \\ &= 7x^2 + x.\end{aligned}$$

When subtracting polynomials, we have to be careful to distribute a negative sign to each term.

Example: Simplify $(5x^3 - 3x^2 + 8x - 7) - (4x^3 + 7x^2 - 15x - 6)$.

Solution: We have

$$\begin{aligned}(5x^3-3x^2+8x-7)-(4x^3+7x^2-15x-6) &= 5x^3-3x^2+8x-7-4x^3-7x^2-(-15x)-(-6)\\ &= 5x^3-3x^2+8x-7-4x^3-7x^2+15x+6\\ &= 5x^3-4x^3+-3x^2-7x^2+8x+15x+-7+6\\ &= x^3-10x^2+23x-1.\end{aligned}$$

Example: Simplify $(x^2y^2-3x^2y+5xy^2)+(3x^2y^2-x^2y-8xy^2)$.

Solution: Keep in mind that we just add the like terms. We have

$$\begin{aligned}&x^2y^2+3x^2y^2-3x^2y-x^2y+5xy^2-8xy^2\\ &=4x^2y^2-4x^2y-3xy^2.\end{aligned}$$

A polynomial is a sum of many terms. The degree of a term is the sum of the exponents of the variables of that term. The degree of a polynomial is the maximum of the degrees of all the terms in that polynomial.

Ex) The polynomial $3x^4-7x^3+x+1$ has a term of degree 4, a term of degree 3, a term of degree 1, and a term of degree zero (note that we can write 1 as $1x^0$). Since the maximum of the degrees of all the terms is 4, this is a fourth-degree polynomial.

Ex) The polynomial $x-1$ is of degree 1, since it has a term of degree 1 and a term of degree 0.

Ex) The polynomial $xy+x^2y-xy^2+y^2-x^2$ has a term of degree 3. The degree of xy is 2, since we add the exponents on x and y. The degree of x^2y is 3, since we add the exponents on x^2 and y. Similarly, the degree of $-xy^2$ is 3, and the degrees of y^2 and $-x^2$ are both 2. The maximum of all these degree terms is 3, so this is a third-degree polynomial.

A polynomial expression with one term is sometimes called a monomial. The root of the word is *mono-*, which means "one." So $3x^2$ is an example of a monomial, and $14x^3a^2$ is an example of a monomial.

A binomial is a two-term polynomial expression. Examples are x^2+y^3 and $2x-1$.

A trinomial is a polynomial expression with three terms. So $x^2 + x + 1$ is a second-degree trinomial.

Multiplication of Polynomials

In multiplication of polynomials we make sure to obey the distributive property. For example, if we multiply $x + 1$ by $x - 1$, we have

$$(x+1)(x-1),$$

so we must be sure that x gets multiplied by $x - 1$ and that 1 gets multiplied by $x - 1$. We have

$$\begin{aligned}(x+1)(x-1) &= x(x-1)+1\cdot(x-1)\\ &= x^2 - x + x - 1\\ &= x^2 - 1.\end{aligned}$$

To remember to do this, some students apply the acronym FOIL, which stands for "First Outer Inner Last." See Figure 5-1.

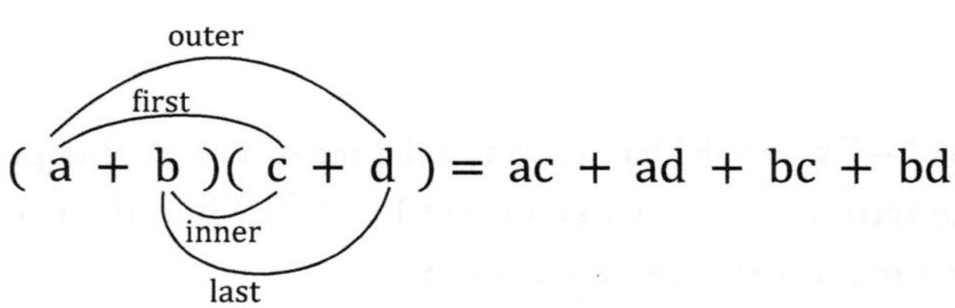

Fig. 5-1

Ex) To multiply $(x + 5)(x - 2)$, we have

$$\begin{aligned}(x+5)(x-2) &= x^2 - 2x + 5x - 10\\ &= x^2 + 3x - 10.\end{aligned}$$

Example: Multiply out $(a + b)^2$.

Solution: The answer is *not* $a^2 + b^2$. Think of $(a+b)^2$ as $(a+b)(a+b)$. After all, being squared is just shorthand for multiplying something by itself. Then we FOIL, with this result:

$$\begin{aligned}(a+b)^2 &= (a+b)(a+b)\\ &= a^2 + ab + ab + b^2\\ &= a^2 + 2ab + b^2.\end{aligned}$$

A good way to remember to FOIL is to think of $(a+b)(a+b)$ as $a+b$ copies of $a+b$. So we need a copies of $a+b$, plus b copies of $a+b$. This is $a(a+b)+b(a+b)$.

Example: Simplify $(2x+3y)^2 - (2x-3y)^2$.

Solution: Remember that we are subtracting, so be careful about simplifying the second expression here. Let's use brackets so we don't forget to distribute that negative after we FOIL.

$$\begin{aligned}(2x+3y)(2x+3y)-(2x-3y)(2x-3y) &= 4x^2 + 6xy + 6xy + 9y^2 - [4x^2 - 6xy - 6xy + 9y^2]\\ &= 4x^2 + 12xy + 9y^2 - [4x^2 - 12xy + 9y^2]\\ &= 4x^2 + 12xy + 9y^2 - 4x^2 + 12xy - 9y^2\\ &= 24xy.\end{aligned}$$

Example: Simplify $(x^3 - x^2 + x + 1)(x^3 + 2x^2 - x - 1)$.

Solution: We have to be systematic about this. We take one term of the first factor, and multiply it by the second, then take the next term, and so on.

So we start with x^3 being multiplied by $(x^3 + 2x^2 - x - 1)$, then $-x^2$ being multiplied by $(x^3 + 2x^2 - x - 1)$, and so on:

$$\begin{aligned}&x^3(x^3+2x^2-x-1) - x^2(x^3+2x^2-x-1) + x(x^3+2x^2-x-1) + 1(x^3+2x^2-x-1)\\ &= x^6 + 2x^5 - x^4 - x^3 - x^5 + 2x^4 - x^3 + x^2 + x^4 + 2x^3 - x^2 - x + x^3 + 2x^2 - x - 1\\ &= x^6 + x^5 + 2x^4 + x^3 + 2x^2 - 2x - 1.\end{aligned}$$

Section 5.1 Exercises

Simplify the following expressions by adding or subtracting.

1. $(x^2 - x + 1) + (x^2 + 2x + 3)$
2. $(x^2 - x + 1) - (x^2 - x + 1)$
3. $(x + y) + (x - y)$
4. $(2x + x^2) - (x^2 - x)$
5. $(3x^2 + 4x + 1) - (x^2 - 4x + 1)$
6. $(2x^2 + 4x + 1) - (1 - x + 2x^2)$
7. $(x^3 + 7x^2 - 7x + 11) - (2x^3 - 3x^2 - 5x + 1)$
8. $(x^2y^2 + 2xy) + (3xy - 5x^2y^2)$
9. $(2xy^2 + 5x^2y) - \left(\frac{1}{2}xy^2 - \frac{1}{2}x^2y\right)$
10. $(3xyz + 6xyz^2) + \left(-\frac{1}{4}xyz^2 + xyz\right)$
11. Give the degree of the polynomial that is the result of Exercise 1.
12. Give the degree of the polynomial that is the result of Exercise 2.
13. Give the degree of the polynomial that is the result of Exercise 3.
14. Give the degree of the polynomial that is the result of Exercise 4.
15. Give the degree of the polynomial that is the result of Exercise 5.
16. Give the degree of the polynomial that is the result of Exercise 6.
17. Give the degree of the polynomial that is the result of Exercise 7.
18. Give the degree of the polynomial that is the result of Exercise 8.
19. Give the degree of the polynomial that is the result of Exercise 9.
20. Give the degree of the polynomial that is the result of Exercise 10.

Multiply out the following expressions, and simplify the result.

21. $x(x + 1)$
22. $(x + 4)(x - 4)$
23. $(x - 8)(x + 8)$
24. $(x + 1)(x - 2)$
25. $(x + 4)(x + 5)$
26. $(x + 2)(x - 5)$
27. $(x + 10)(x - 11)$
28. $(x + 15)(x - 1)$
29. $\left(x + \frac{1}{2}\right)\left(x + \frac{3}{2}\right)$

30. $(2x+1)(x+1)$
31. $(3x+1)(2x+3)$
32. $(3x-5)(2x-7)$
33. $(2x+1)(x-5)$
34. $(4x+7)(8x-5)$
35. $(3x+10)(2x-9)$
36. $(x+2)^2$
37. $(x-2)^2$
38. $(x+1)^2$
39. $(x-1)^2$
40. $(2x+3)^2$
41. $(3x-5)^2$
42. $(x^2+x+1)(x^2-x-1)$
43. $(3x^2+x-1)(x^2+2x+2)$
44. $(2x^2+x+1)(x^2+3x+5)$
45. $(x+1)^3$
46. $(x-1)^3$

5.2: Polynomial Long-Division

The division of polynomials is a very sensitive business; we have to be careful not to break any basic rules. In this section, we will cover long division and synthetic division. But first, note that dividing expressions by a single term is pretty easy to do. We can split the fraction up.

Ex) If we have $\frac{3x+1}{2}$, we can write this as $\frac{3x}{2} + \frac{1}{2}$.

If we have a variable, just be careful to remember the quotient rule for exponents:

$$\frac{a^m}{a^n} = a^{m-n}$$

Example: Simplify $\frac{x^4-4x^3+8x^2}{2x}$.

Solution: We split the fraction up for

$$\frac{x^4 - 4x^3 + 8x^2}{2x} = \frac{x^4}{2x} - \frac{4x^3}{2x} + \frac{8x^2}{2x}$$
$$= \frac{x^3}{2} - 2x^2 + 4x.$$

Example: Simplify $(3x^5 - 9x^4) \div (-3x^2)$.

Solution: Splitting the fraction because the denominator is just one term, we have

$$\frac{3x^5}{-3x^2} + \frac{-9x^4}{-3x^2}$$
$$= -x^3 + 3x^2.$$

Now, it is critical to remember that this cannot be done when there is more than one term in the denominator. So

$$\frac{1}{x+1} \neq \frac{1}{x} + \frac{1}{1}.$$

The reason we can split fractions when there is just one term in the denominator is that we are going the opposite way in adding fractions:

$$\frac{a+b}{c} = \frac{a}{c} + \frac{b}{c}.$$

If we go from right to left in the previous equation, both fractions have the same denominator so we just add the numerators.

Ex) We can write $\frac{x+1}{x}$ as

$$\frac{x+1}{x} = \frac{x}{x} + \frac{1}{x} = 1 + \frac{1}{x},$$

as long as $x \neq 0$. But note that

$$\frac{x}{x+1}$$

cannot be manipulated in the same manner.

Long Division

If there are more terms in the denominator, we must use long division. To do this, it helps to write out long division the way we used to do it with numbers. For example, to long-divide 3 by 2, we do the following.

1. First, we **divide** to get 1, and we write this:

$$\begin{array}{r} 1 \\ 2\overline{)3} \end{array}$$

2. Next, we **multiply** the 1 and 2:

$$\begin{array}{r} 1 \\ 2\overline{)3} \\ 2 \end{array}$$

3. Then, we **subtract**:

$$\begin{array}{r} 1 \\ 2\overline{)3} \\ \underline{-2} \\ 1 \end{array}$$

4. The last step is to **bring down**, but here there is nothing more to bring down.

Now, since 2 does not evenly divide into 1, this is the remainder, so we put a plus at the top. This is followed by the remainder divided by 2:

$$\begin{array}{r} 1+\frac{1}{2} \\ 2\overline{)\;3\;\;} \\ -2 \\ 1 \end{array}$$

The same steps are done in polynomial long division:

1. **Divide.** Take the first term of the divisor and see how many times it will go into the dividend. Often we can just put the first term of the dividend divided by the first term of the divisor to get this. Write this up top.
2. **Multiply** the result by the entire divisor, and write it below the dividend.
3. **Subtract.** It is easiest to change all of the signs of the polynomial we wrote so that we can just add a negative.
4. **Bring down** any remaining terms so we don't forget them, and repeat the process.

Example: Perform the division $(7x^4+3x^3+x^2-10x+5)\div(x^2+1)$.

Solution: Setting up the division, we have

$$x^2+1\overline{)7x^4+3x^3+x^2-10x+5}$$

The first step is to divide. How many times does x^2 go into $7x^4$? We can write this as

$$\frac{7x^4}{x^2}$$

and see that the result is $7x^2$. Then we multiply $7x^2(x^2+1)$ to write $7x^4+7x^2$ below:

$$\begin{array}{rl} & \qquad\qquad\qquad\qquad 7x^2 \\ x^2+1 & \overline{)7x^4+3x^3+x^2-10x+5} \\ & \;\; 7x^4+\qquad 7x^2 \end{array}$$

Now it is helpful to switch the signs of that $7x^4 + 7x^2$, because we need to subtract:

$$
\begin{array}{r}
7x^2 \\
x^2+1\,\big)\overline{7x^4 + 3x^3 + x^2 - 10x + 5} \\
\underline{-7x^4 \qquad - 7x^2} \\
-6x^2 \qquad\qquad\;\;
\end{array}
$$

Let's not forget to bring down everything else:

$$
\begin{array}{r}
7x^2 \\
x^2+1\,\big)\overline{7x^4 + 3x^3 + x^2 - 10x + 5} \\
\underline{-7x^4 \qquad -7x^2} \qquad\qquad\;\; \\
3x^3 - 6x^2 - 10x + 5
\end{array}
$$

Now we repeat the process. How many times does x^2 go into $3x^3$? It goes in $3x$ times, because

$$\frac{3x^3}{x^2} = 3x.$$

Below is the finished process:

$$
\begin{array}{r}
7x^2 + 3x - 6 + \dfrac{-13x + 11}{x^2 + 1} \\
x^2+1\,\big)\overline{7x^4 + 3x^3 + x^2 - 10x + 5} \\
\underline{-7x^4 \qquad -7x^2} \qquad\qquad\;\; \\
3x^3 - 6x^2 - 10x + 5 \\
\underline{-3x^3 \qquad -3x} \qquad\; \\
-6x^2 - 13x + 5 \\
\underline{+6x^2 \qquad +6} \\
-13x + 11
\end{array}
$$

We know we have finished because x^2 does not evenly divide $-13x$. We would need at least an x^2 to be able to divide again, so when the degree of what we have is less than the degree of the divisor, the process is complete.

Example: Perform the division

$$(6x^4 + 7x^3 - 29x^2 - 34x - 6) \div (3x^2 - 4x - 6).$$

Solution: We put $3x^2 - 4x - 6$ on the outside, and we have

$$3x^2 - 4x - 6\overline{\smash{)}6x^4 + 7x^3 - 29x^2 - 34x - 6}$$

Next, we see that

$$\frac{6x^4}{3x^2} = 2x^2,$$

so the first part of our quotient is $2x^2$. We perform the rest of the operation as shown below:

$$\begin{array}{r}
2x^2 + 5x + 1 \\
3x^2 - 4x - 6\overline{\smash{)}6x^4 + 7x^3 - 29x^2 - 34x - 6} \\
\underline{-6x^4 + 8x^3 + 12x^2} \qquad\qquad \\
15x^3 - 17x^2 - 34x - 6 \\
\underline{-15x^3 + 20x^2 + 30x} \quad \\
3x^2 - 4x - 6 \\
\underline{-3x^2 + 4x + 6} \\
0 \quad
\end{array}$$

Note that the result shows that $3x^2 - 4x - 6$ divides evenly into $6x^4 + 7x^3 - 29x^2 - 34x - 6$. The two quadratic expressions $3x^2 - 4x - 6$ and $2x^2 + 5x + 1$ multiply together to give $6x^4 + 7x^3 - 29x^2 - 34x - 6$. In other words,

$$(3x^2 - 4x - 6)(2x^2 + 5x + 1) = 6x^4 + 7x^3 - 29x^2 - 34x - 6.$$

Synthetic Division

We can do synthetic division when we're dividing a polynomial by a first-degree polynomial.

This is sometimes preferred, because instead of dividing and subtracting, as in long division, synthetic division involves multiplying and adding.

For example, suppose we are dividing $x^3 - 5x^2 + 8x - 1$ by $x - 2$. We set the $x - 2$ equal to zero:

$$x - 2 = 0.$$

So we synthetically divide by the number 2. We put this number in a box, and then write the coefficients of the dividend next to it. We bring the first coefficient down, and then multiply the boxed number by this number. The result goes below the second dividend coefficient. Then we add, and multiply, and keep repeating this process until we have no more dividend coefficients. The remainder will be the final number obtained.

We have

$$\begin{array}{r|rrrr} \boxed{2} & 1 & -5 & 8 & -1 \\ & \downarrow & 2 & -6 & 4 \\ \hline & 1 & -3 & 2 & |3 \end{array}$$

The last number written is a 3, and this is the remainder. To get the quotient, the numbers in the bottom row are the coefficients. We use a polynomial that is one degree lower than what we divided. Since we were dividing a third-degree polynomial, the quotient will be a second-degree polynomial $ax^2 + bx + c$. The coefficients $1, -3, 2$ are taken from the bottom row. So we have

$$(x^3 - 5x^2 + 8x - 1) \div (x - 2) = x^2 - 3x + 2 + \frac{3}{x - 2}.$$

Example: Perform the division

$$\left(10x^5 - \frac{1}{2}x^3 + x^2 - 7x + 1\right) \div (2x + 1).$$

Solution: First, we set $2x+1=0$ to see that we should be synthetically dividing by $-\frac{1}{2}$. This goes in our box, and we write the coefficients of $10x^5-\frac{1}{2}x^3+x^2-7x+1$ next to it. We have to leave a 0 as a placeholder for $0x^4$. The synthetic division handles each power separately, so if one of them is missing in the dividend we leave a zero for the coefficient. Our setup is this:

$$\begin{array}{r|rrrrrr} \boxed{-\frac{1}{2}} & 10 & 0 & -\frac{1}{2} & 1 & -7 & 1 \\ & \downarrow & & & & & \\ \hline \end{array}$$

We bring the 10 down, and multiply by $-\frac{1}{2}$ to get -5, which is written below the 0. Then we add 0 to -5 to get -5, and write that below the bar. Then we multiply by $-\frac{1}{2}$ and get $\frac{5}{2}$, and write that below the coefficient $-\frac{1}{2}$, and so on. The result, shown below, has a separating bar for the last number because that is the remainder.

$$\begin{array}{r|rrrrrr} \boxed{-\frac{1}{2}} & 10 & 0 & -\frac{1}{2} & 1 & -7 & 1 \\ & \downarrow & -5 & \frac{5}{2} & -1 & 0 & \frac{7}{2} \\ \hline & 10 & -5 & 2 & 0 & -7 & |\,\frac{9}{2} \end{array}$$

Now, the coefficients of the polynomial one degree lower than what we started with are $10x^4-5x^3+2x^2-7$, and the remainder is $\frac{9}{2}$. But our answer is not complete yet.

When we synthetically divide by $-\frac{1}{2}$, we're dividing by $x+\frac{1}{2}$. But that's not what we started with. We can think of this as factoring the 2 out of the denominator of the original problem:

$$\frac{10x^5-\frac{1}{2}x^3+x^2-7x+1}{2x+1}=\frac{10x^5-\frac{1}{2}x^3+x^2-7x+1}{2\left(x+\frac{1}{2}\right)}.$$

So whatever we synthetically divide by $-\frac{1}{2}$, we are dividing by $x+\frac{1}{2}$. We need to be careful to take the 2 that we factored out in the denominator and carry it all the way to the quotient:

$$\frac{10x^5-\frac{1}{2}x^3+x^2-7x+1}{2\left(x+\frac{1}{2}\right)}=\frac{1}{2}\left[\frac{10x^5-\frac{1}{2}x^3+x^2-7x+1}{\left(x+\frac{1}{2}\right)}\right]$$
$$=\frac{1}{2}\left[10x^4-5x^3+2x^2-7+\frac{\frac{9}{2}}{x+\frac{1}{2}}\right].$$

This is why the quotient and the remainder are both multiplied by $\frac{1}{2}$. Our quotient is

$$5x^4-\frac{5}{2}x^3+x^2-\frac{7}{2}+\frac{\frac{9}{2}}{2x+1}.$$

This last step—carrying the $\frac{1}{2}$ through the division—does not need to be done every time. We just have to remember to take our polynomial one degree lower and multiply it by something. The easy way to remember this is that if we divide by $ax+b$, we synthetically divide by $-\frac{b}{a}$; but we need to remember to multiply what we get by $\frac{1}{a}$.

Example: Divide

$$\frac{6x^4-17x^3+35x^2-22x+5}{3x-1}.$$

Solution: Since we are dividing by a linear factor, we can use synthetic division. We set $3x-1=0$ to see that we should synthetically divide by $\frac{1}{3}$. The setup is as follows:

$$\begin{array}{c|ccccc} \boxed{\frac{1}{3}} & 6 & -17 & 35 & -22 & 5 \\ & \downarrow & & & & \\ \hline & & & & & \end{array}$$

Once we bring down the 6, we multiply, then add, then multiply, then add, and so on. The finished calculation is given below:

$$\begin{array}{c|ccccc} \boxed{\frac{1}{3}} & 6 & -17 & 35 & -22 & 5 \\ & \downarrow & 2 & -5 & 10 & -4 \\ \hline & 6 & -15 & 30 & -12 & |1 \end{array}$$

We obtain $6x^3 - 15x^2 + 30x - 12 + \frac{1}{x - \frac{1}{3}}$, but we must remember to multiply this result by $\frac{1}{3}$. Our quotient is

$$\frac{1}{3}\left(6x^3 - 15x^2 + 30x - 12 + \frac{1}{x - \frac{1}{3}}\right)$$
$$= 2x^3 - 5x^2 + 10x - 4 + \frac{1}{3x - 1}.$$

In the end, synthetic division is easier, but it is so much more important to know long division. Long division will work when we're dividing by a polynomial of any degree. Synthetic division works only when we're dividing by $ax + b$.

Section 5.2 Exercises

1. Simplify

$$\frac{x^2 + x}{x}.$$

2. Simplify

$$\left(3x^2 + x^4\right) \div x^2.$$

3. Simplify $(40x^3 - 15x^2) \div (5x)$.
4. Simplify $(x^{10} - x^7) \div x^2$.

In the following exercises, use long division, or synthetic division if possible. Make sure to indicate what the quotient is (especially if using synthetic division) and any remainder.

5. $(x^4 + x^3 - 2x^2 + x - 1) \div (x^2 + 1)$
6. $(x^2 + x - 1) \div (x + 1)$
7. $(2x^2 + x + 7) \div (x - 5)$
8. $(3x^3 + x^2 - x + 1) \div (x + 2)$
9. $(6x^3 + 3x^2 - 5x + 10) \div (3x + 2)$
10. $(3x^3 - 5x^2 + 10x + 6) \div (2x - 1)$
11. $(x^4 - x^2 + 1) \div (x^2 + 1)$
12. $(3x^2 + 2x + 5) \div (x + 5)$
13. $(x^4 - x^3 + 2x^2 + 6x - 5) \div (x^3 + x + 2)$
14. $(2x^4 - 5x^3 + 7x^2 + 10x - 5) \div (x^2 - 2x + 1)$
15. $(3x^5 - 2x^4 - x^3 + x^2 + 2x - 1) \div (x^3 - 2x^2 + 2x - 1)$
16. $(3x^4 - x + 1) \div (x^2 + 1)$
17. $(2x^4 - 5x^3 - x^2 + x + 1) \div (x - 1)$
18. $(x^4 - 10x^3 + x^2 - 6x - 5) \div (x - 2)$
19. $(3x^5 - 4x^4 + 10x^3 - 7x^2 + 10x - 5) \div (x + 1)$
20. $(2x^4 - x^2 + x) \div (2x - 1)$
21. $(2x^4 - 3x^3 + 7x^2 - x + 1) \div (3x + 2)$

5.3: Factoring by GCF and Factoring by Grouping

In this section we will discuss the beginnings of factoring, which is one of the most important simplifications to learn in mathematics. We make liberal use of the distributive property here, which says that

$$a(b + c) = ab + ac.$$

But what we will be doing is going in the reverse direction. That is,

$$ab + ac = a(b + c).$$

For example, when we write $2x+2$, we sometimes simplify this by factoring out the 2. In the distributive property above, $a = 2$. So going backwards, we factor the 2 out to get

$$2x+2=2(x+1).$$

Factoring is important for many reasons. Factored computations are more simplified, and they make things easier to work with. The factors of something are easier to deal with than the multiplied-out long expression.

Recall Section 5.1, where we were multiplying out these expressions and asking "Why did we do that?" Well, factoring is actually more important than multiplying things out, but it is much harder to do. So we needed the practice building our intuition. For example, we know what $(a+b)^2$ looks like when it's multiplied out. If we are asked to factor $a^2 +2ab+b^2$, then we have an idea of what it should be.

Greatest Common Factor

If we have a complicated expression, we need to look at each term and decide the greatest common factor. The greatest common factor, or GCF, is the largest factor common to all terms.

Ex) The GCF for the terms of $x^2 +2x$ is x.

Ex) The GCF for the terms of the expression $x^2 +2x+1$ is 1.

Ex) The GCF of $3x^2 +6x$ is $3x$.

The procedure is to look at the constants, and then look at the variables one by one. We take out the lowest exponent.

Example: Factor $3a^3 -9a^2$.

Solution: One term has a 3, and the other term has a 9, which is 3^2, so the most we can take out is 3 (the smaller of the two factors). Likewise, since one term has an a^3 and the other term has a^2, we can take out a^2 since it is the largest amount common to both. Notice that the largest amount common is also the smallest exponent. So we have

$$3a^3-9a^2=3a^2(a-3).$$

No matter how many variables we have, we go term-by-term in the expression to find what is common to all terms.

Ex) In the expression $27x^3y^2 + 9xy^2z + 6x^4y^3$, let's look at the coefficients first. They are 27, 9, and 6. Writing them in factored form, we have 3^3, 3^2, and 2.3. Since we can take out only the smallest amount of what's there, we can take out only the 3 that is common to all terms. Now let's look at the powers of x. We have x^3, x, and x^4. Since we can only take out the smallest amount, we can bring out an x. And then on to y, we have exponents of 2, 2, and 3. So we can bring out a y^2 from each term. And there is only one term with a z, so we cannot take it out. So our greatest common factor is $3xy^2$. Now, when we take $3xy^2$ out from each term, we have to determine what's left. One way to think about this is to just determine what's left over. If we take out $3xy^2$ from $27x^3y^2$, a factor of 9 and a factor of x^2 will be left over. Similarly, we can do the rest and obtain

$$27x^3y^2 + 9xy^2z + 6x^4y^3 = 3xy^2(9x^2 + 3z + 2x^3y).$$

Two things to be aware of: First, once we factor out the GCF, none of the terms left inside the parentheses should have factors all in common. If we left something in common to all terms, we just factor that out as well. Second, we can always check that we factored correctly, by multiplying everything back out again.

YOU TRY IT: Multiply out $3xy^2(9x^2 + 3z + 2x^3y)$ to check that it equals $27x^3y^2 + 9xy^2z + 6x^4y^3$.

Another way to determine what's left is to actually do the division. If we take out a factor of $3xy^2$ from $27x^3y^2$, what's left over is

$$\frac{27x^3y^2}{3xy^2},$$

which simplifies to $9x^2$. This is the same as above, so we can continue in this manner if necessary. This will be helpful to us in Section 7.2 when the exponents become more complicated.

Example: Factor $4ab + 5b^2 + 21a^2b$.

Solution: Look at the coefficients 4, 5, and 21. These numbers have no factors in common. If we don't see this, we can always prime-factor them as 2^2, 5, and 3×7.

Next, we have only b in common for the other terms. While a is common to the first and third terms, we cannot factor it out of the entire expression. So our GCF is b.

Thus,

$$4ab + 5b^2 + 21a^2b = b(4a + 5b + 21a^2).$$

Again, a convenient way to check the work is to note that there is nothing in common to the terms inside the parentheses.

Example: Factor $4a^2(y+z) - 34b(y+z)$.

Solution: We might be tempted to multiply out here. But notice that $(y + z)$ is common to both terms, so it is part of the GCF. Now, the first term has a coefficient of 4 and the second term has coefficient of 34, so we can take out a 2 (since this is 2^2 and 2×17). The result is

$$2(y+z)(2a-17b).$$

Factoring by Grouping

Factoring by grouping is a special factoring method that works when there are four terms. If we factor the first pair of terms and the second pair of terms, we want to discover if those pairs have anything in common that can be factored out.

Ex) If we are factoring $x^3 + x^2 + x + 1$, consider the first two terms $x^3 + x^2$. If we factor out x^2, we end up with $x^2(x+1)$. Now look at the second pair of terms, $x + 1$. This cannot be factored. But our pairs now have $(x + 1)$ in common, so *that* can be factored out! It works as follows:

$$\begin{aligned} x^3 + x^2 + x + 1 &= x^2(x+1) + (x+1) \\ &= (x+1)[x^2+1] \\ &= (x^2+1)(x+1). \end{aligned}$$

Example: Factor $x^4 + xy^2 + x^3y + y^3$.

Solution: Let's think about why we could not factor this until now. There is nothing common to each term, so we can't take out a GCF *for all four terms*. The giveaway is that there are four terms, so let's try grouping. The GCF of the first two terms is x, and the GCF of the second pair of terms is y:

$$x^4 + xy^2 + x^3y + y^3 = x(x^3 + y^2) + y(x^3 + y^2).$$

Now notice that the expression $(x^3 + y^2)$ is common to both x and y, so we factor out that whole term:

$$(x^3 + y^2)[x + y].$$

The result is $x^4 + xy^2 + x^3y + y^3 = (x^3 + y^2)(x + y)$.

Example: Factor $3ax - 3bx - 2a + 2b$ by grouping.

Solution: The first two terms, $3ax$ and $-3bx$, have $3x$ in common. The third and fourth terms, $-2a$ and $2b$, have the number 2 in common. So factoring in pairs, we have

$$3ax - 3bx - 2a + 2b = 3x(a - b) + 2(-a + b).$$

Note that the binomials in parentheses are not equal, so they cannot immediately be factored out. However, they are close. In fact, if we had factored out -2 we would be fine. To match up the expressions in parentheses, we need to factor out an extra -1. Note that

$$-a + b = -(a - b).$$

So we have

$$\begin{aligned} 3x(a - b) + 2(-a + b) &= 3x(a - b) - 2(a - b) \\ &= (a - b)[3x - 2] \\ &= (a - b)(3x - 2). \end{aligned}$$

Section 5.3 Exercises

Factor the following expressions.

1. $xy - x$
2. $60x + 48y + 6z$
3. $x^2y - xy$
4. $x - x^2$
5. $2x^2 + 3x^3 + 6x^4$
6. $14x + 10x^2$
7. $64x^3 - 70x^5$
8. $18a^2 - 9a$
9. $80xy + 10xy^2$
10. $3x^3 + 9x$
11. $50b^2 + 10b - 5$
12. $13y^4x^{10} - 39x^{11}y^3$
13. $6x^2y^3 - 8xy^4 + 14x^3y^2$
14. $15a^3b^6 + 35a^2b^2 + 65a^5b^4$
15. $11x^2y^4 + 121x^3y^5$
16. $3x^3y^7 - x^2y^9$
17. $x^4 + 10x^3 + x^2$
18. $x(a^2 + b^2) + 2y(a^2 + b^2)$

Factor by grouping.

19. $x^3 + x^2 + 4x + 4$
20. $ay + by + 6a + 6b$
21. $2x + 12 + bx + 6b$
22. $abc + ab - c^2 - c$
23. $ab^2 + b^3 - a - b$
24. $x^3 + x^2y - 7x - 7y$
25. $x^4 + x^3 + xy + y$
26. $6x^3y + 9xy + 14x^2 + 21$
27. $8ab + 2ac - 28b^2 - 7bc$
28. $ab + bd + ac + cd$

5.4: Factoring Trinomials

A quadratic expression has the form $ax^2 + bx + c$. This expression will be used so much in this and future sections that each piece of it has a name:

- ax^2 is called the quadratic term,
- a is called the coefficient of the quadratic term,
- bx is called the linear term,
- b is called the coefficient of the linear term, and
- c is called the constant.

Quadratic Coefficients Equal to 1

Factoring quadratic expressions is one of the most important topics in algebra, because it can appear in any future math course.

Some quadratic expressions can be factored into two binomials $(ax+b)(cx+d)$. This section discusses how this is done. First, consider the case when the quadratic coefficient is 1. We would factor this as $(x+b)(x+d)$. If we FOIL out $(x+b)(x+d)$, we have

$$x^2 + (b+d)x + bd.$$

The constant term is bd, and the coefficient of the linear term is $b+d$. To figure out what these numbers have to be, we multiply them together to get the constant term and we add them together to get the linear coefficient.

Example: Factor $x^2 + 2x + 1$.

We factor this into a product of two binomials. The coefficient of the quadratic term is 1, so our factors are

$$(x+?)(x+?).$$

The constant in $x^2 + 2x + 1$ is 1, and the coefficient of the linear term is 2, so we need two numbers that multiply together to get 1 but add up to 2. Well, both numbers being 1 will work:

$$x^2 + 2x + 1 = (x+1)(x+1).$$

Remember we can always check by FOILing out the right-hand side.

Note that this is called a perfect square trinomial, because we can write this as a product of the same binomial: $x^2 + 2x + 1 = (x+1)(x+1) = (x+1)^2$.

Example: Factor $x^2 - 13x + 42$.

Solution: The coefficient of the quadratic term is 1, so we factor this as

$$(x + ?)(x + ?).$$

Now, we need two factors of 42 that will add up to −13. If we write out the factor pairs of 42, we have

$$\begin{array}{c} 1 \cdot 42 \\ 2 \cdot 21 \\ 3 \cdot 14 \\ 6 \cdot 7. \end{array}$$

The last pair will work, because $-6x$ and $-7x$ add up to $-13x$ in the FOILed product. So we have

$$x^2 - 13x + 42 = (x-6)(x-7).$$

Example: Factor $x^2 - 2x - 24$.

Solution: The coefficient of the quadratic term is 1, so we just need two numbers that multiply together to get –24 and add together to get –2. The numbers that multiply together to get 24 are

$$1 \cdot 24$$
$$2 \cdot 12$$
$$3 \cdot 8$$
$$4 \cdot 6$$

Note that a +4 and a –6 adds up to –2 and multiplies together to get –24. So our factorization is

$$x^2 - 2x - 24 = (x+4)(x-6).$$

Quadratic Coefficients Not Equal to 1

If the quadratic coefficient a is a prime number (it has only itself and 1 as a factor), then we have to factor $ax^2 + bx + c$ into $(ax+?)(x+?)$, because that's the only way the quadratic term will have a coefficient of a.

Example: Factor $2x^2 - 5x - 12$.

Solution: The coefficient of the quadratic term is 2, so we have to start our factors with $2x$ and x. The order won't matter here; we could just as easily do x and $2x$. So we have

$$(2x+?)(x+?).$$

Now we need the numbers that multiply up to get 12. They are 1 and 12, 2 and 6, and 3 and 4. Note that we cannot just look for those pairs to get to 5 (as, indeed, none of them will get us that). So we use trial-and-error to see which pair will work. If we do

$$(2x\quad 6)(x\quad 2),$$

there are no signs (plus or minus) that we can insert here that will get us a $-5x$. But if we try

$$(2x\quad 3)(x\quad 4),$$

we see that making the 4 negative and the 3 positive will give a cross term of $-8x+3x$ when we FOIL this out. So our factorization is

$$2x^2-5x-12=(2x+3)(x-4).$$

If the coefficient of the quadratic term is not prime (a natural number greater than 1 that is not prime is called composite), then the trial-and-error approach becomes a little more difficult. However, the more we practice this, the easier it becomes.

Example: Factor $12x^2+17x-5$.

Solution: The coefficient of the quadratic term is 12, so first we need coefficients of x to multiply together to get $12x^2$. Our choices are x and $12x$, $2x$ and $6x$, or $3x$ and $4x$. We can't pick the pair with 12 or the one with 6, because that will be too high for a constant term of 5. So let's try $3x$ and $4x$:

$$(3x+?)(4x+?).$$

As for the factors of 5, our only choices are 5 and 1. If we pick a +5 and −1, we have

$$(3x+5)(4x-1),$$

which will give us $12x^2+20x-3x-5$, which is $12x^2+17x-5$ as required.

The *ac*-Method

Another method for factoring trinomials—the *ac*-method—depends a lot less on trial and error. In the *ac*-method, we take the product of the coefficient of the quadratic term and the constant, which is ac in the quadratic expression ax^2+bx+c. Then, we find two numbers that add up to ac. Finally, we add the factors together and we factor by grouping.

For example, let's consider the factors $(3x+1)(7x-4)$. If we multiply this out, we can write it as $3x(7x-4)+1(7x-4)$, which could be taken one more step to be

$$21x^2-12x+7x-4.$$

Note here that $21\cdot(-4)$ is -84, which is the same as the product of -12 and 7. So if we can find two numbers (-12 and 7) that multiply together to get the same thing as $a\cdot c$ and that add together to get the same number as the coefficient of the linear term, then we will be able to use factoring by grouping.

Example: Factor $10x^2+11x-6$ using the *ac*-method.

Solution: The product of the coefficient of the quadratic term and the constant is $10(-6)=-60$. So we need to find numbers that multiply together to get -60 and add together to get 11. For 60, our possible pairs are

$$1\cdot 60,\quad 2\cdot 30,\quad 3\cdot 20,\quad 4\cdot 15,\quad 5\cdot 12,\quad 6\cdot 10.$$

The only way to get 11 out of these pairs is to use 4 and 15. We will need $15x-4x$ to get $11x$. We write our quadratic as

$$10x^2+11x-6=10x^2+15x-4x-6.$$

Now we should be able to use factoring by grouping. We must make sure not to add the $15x$ and the $-4x$ again, or we'll be going backwards. The pair $10x^2+15x$ has a $5x$ that can be factored out, and the $-4x-6$ has a common factor of -2. So we have

$$\begin{aligned}10x^2+11x-6&=10x^2+15x-4x-6\\&=5x(2x+3)-2(2x+3)\\&=(2x+3)[5x-2]\\&=(2x+3)(5x-2).\end{aligned}$$

Some students prefer the *ac*-method because there is less trial-and-error than what we talked about earlier. Give both methods a try, and see which you prefer.

Section 5.4 Exercises

Factor the following trinomials as the product of two binomials.

1. $x^2 + 3x + 2$
2. $x^2 + x - 2$
3. $x^2 - x + 2$
4. $x^2 - 3x + 2$
5. $x^2 + x - 30$
6. $x^2 - 4x - 45$
7. $x^2 - 3x - 28$
8. $x^2 + 5x - 6$
9. $x^2 - x - 6$
10. $x^2 + x - 42$
11. $x^2 - 2x - 8$
12. $x^2 + 21x + 110$
13. $x^2 - x - 56$
14. $x^2 - 6x - 72$
15. $x^2 + 3x - 40$
16. $2x^2 + 3x + 1$
17. $2x^2 + 5x - 3$
18. $3x^2 - 13x - 10$
19. $3x^2 + 5x - 2$
20. $7x^2 + 6x - 1$
21. $5x^2 - 34x - 7$
22. $5x^2 - 8x - 2$
23. $9x^2 - 17x - 2$
24. $4x^2 + 4x - 3$
25. $6x^2 + 7x + 2$
26. $15x^2 + 13x + 2$
27. $8x^2 + 14x + 3$
28. $8x^2 + 2x - 3$
29. $8x^2 + 6x - 5$
30. $6x^2 - 5x - 56$
31. $12x^2 - 4x - 21$

5.5: Factoring Completely and Difference of Squares

This section will discuss a few more factoring patterns that are common appearances. The first is the difference of squares. This is the special factoring pattern for $a^2 - b^2$:

$$\boxed{a^2 - b^2 = (a+b)(a-b)}$$

This is important because it consists of two binomials where the cross terms cancel. If we multiply out $(a+b)(a-b)$, we have $a^2 + ab - ab + b^2$, which is $a^2 - b^2$.

Ex) $x^2 - y^2$ can be factored as $(x+y)(x-y)$.

Example: Factor $9x^2 - 1$.

Solution: This is a difference of squares. In order to see this, we need to make sure to write the first term as $(3x)^2$, because squaring that out gives $3^2x^2 = 9x^2$. So our factorization is

$$9x^2 - 1 = (3x)^2 - 1^2 = (3x+1)(3x-1).$$

Example: Factor $64x^2 - 81y^2$.

Solution: The first term can be written as $(8x)^2$, and the second term can be written as $(9y)^2$, because $(8x)^2 - (9y)^2$ is the same as $64x^2 - 81y^2$. This is important because to factor $a^2 - b^2$, we need to know just a and b, not their squares. We factor this as follows:

$$64x^2 - 81y^2 = (8x)^2 - (9y)^2 = (8x+9y)(8x-9y).$$

We say a polynomial is <u>prime</u> when it cannot be factored over the real numbers anymore. Another term for this is <u>irreducible</u>. For example, we cannot factor $x^2 + 1$, so we say that $x^2 + 1$ is prime. However, $x^2 - x$ is not prime, because we can factor out an x.

Sometimes it's hard to determine whether a polynomial is irreducible.

Ex) $x^2 + x + 5$ is prime, because there is no pair of numbers that multiply together to get 5 and add together to get 1.

So how can we know that factoring does not work? We'll learn something about quadratics in the future that will help us to know, but for now we just have to determine it based on the practice we've had.

A polynomial is factored completely when it has been written as the product of irreducible polynomials.

Example: Factor $64x^2 - 36y^2$ completely.

Solution: There is a common factor of 4 in both coefficients, so we can write this as

$$64x^2 - 36y^2 = 4(16x^2 - 9y^2).$$

Of course, $16x^2 - 9y^2$ is a difference of squares, so we can factor this completely as

$$4(4x+3y)(4x-3y).$$

Example: Factor $x^4 - 1$ completely.

Solution: Consider $x^4 - 1$ as being $(x^2)^2 - 1^2$. So this is a difference of squares, and we can factor it as

$$x^4 - 1 = (x^2 + 1)(x^2 - 1).$$

We are not finished, though, because $x^2 - 1$ is also a difference of squares! So we have

$$x^4 - 1 = (x^2 + 1)(x^2 - 1) = (x^2 + 1)(x+1)(x-1).$$

Remember that there is no factor pattern for a sum of squares, so this is now factored completely.

Example: Factor $x^3 - x^2 - 2x$ completely.

Solution: Always look for a common factor first. In this case, we can take an x out of every term, so we have

$$x(x^2 - x - 2).$$

Note, however, that we can factor $x^2 - x - 2$ further into a product of binomials. We need numbers that multiply together to get −2 and add together to get −1. Since −2 and 1 will work, we have

$$x^3 - x^2 - 2x = x(x-2)(x+1).$$

Sum and Difference of Cubes

The last factoring patterns we will discuss in this chapter are the sum of cubes and the difference of cubes.

These are their formulas:

$$\boxed{a^3 + b^3 = (a+b)(a^2 - ab + b^2)} \quad \boxed{a^3 - b^3 = (a-b)(a^2 + ab + b^2)}$$

Example: Factor $x^3 + 64$ completely.

Solution: Note that $4^3 = 64$, since $4 \cdot 4 \cdot 4 = 64$. Using the factoring pattern, we factor $(x)^3 + 4^3$ as

$$x^3 + 64 = (x+4)(x^2 - 4x + 16).$$

Example: Factor $d^3 - 8$.

Solution: This is a difference of cubes. We can write $d^3 - 8$ as $(d)^3 - 2^3$. So following the factoring pattern for the difference of cubes gives us

$$d^3 - 8 = (d-2)(d^2 + 2d + 4).$$

Example: Factor $27a^3 - 125b^6$.

Solution: Note that $3^3 = 27$ and $5^3 = 125$. The first term is $(3a)^3$, because this is $3^3 a^3 = 27a^3$. The second term is $(5b^2)^3$, because $(5b^2)^3 = 5^3(b^2)^3 = 125b^6$ (remember that an exponent raised to an exponent means multiplication). So our factorization follows the pattern of the difference of cubes, and we have

$$\begin{aligned} 27a^3 - 125b^6 &= (3a)^3 - (5b^2)^3 = (3a - 5b^2)\left((3a^3)^2 + (3a)(5b) + (5b)^2\right) \\ &= (3a - 5b^2)(9a^6 + 15ab + 25b^2). \end{aligned}$$

It's important to note that the quadratic factor we get from the sum or difference of cubes will most likely be irreducible. So don't worry about factoring it.

Section 5.5 Exercises

Factor the following polynomials, make sure the GCF is factored out, and then use the difference of squares or sum/difference of cubes patterns to completely factor.

1. $x^2 - b^2$
2. $x^2 - 1$
3. $4x^2 - 1$
4. $4x^2 - 9$
5. $x^2 - 4$
6. $4 - x^2$
7. $64w^2 - 81$
8. $49p^2 - 169$
9. $36t^2 - 100$
10. $81x^2 - 25y^2$
11. $121x^2 - 100$
12. $100x^2y^2 - 25$
13. $200y^2 - 98$
14. $x^3 - x$
15. $32x^2 - 18x^4$
16. $36t^3 - 25t$
17. $3x^2 - 27$
18. $x^6 - x^8$
19. $4x^2y^2 - 9y^2$
20. $81x^2a - 100a$

21. $p^3 - q^3$
22. $8x^3 - 1$
23. $t^3 + 27$
24. $n^3 + 64$
25. $8x^3 - 27y^3$
26. $4x^3 - 32$
27. $x^5 + x^8$
28. $27x^3y^3 - 64$
29. $x^8 - 1$

Factor the following polynomials completely. If a polynomial is irreducible, say so.

30. $x^2 + 4$
31. $x^3 + x$
32. $y^3 + y^2 + y$
33. $5x^2 + x + 1$
34. $x^2 - 8x + 16$
35. $2x^2 + 8$
36. $x^3 + 8x^2$
37. $7x^2 - x + 1$
38. $x^2 - 8x + 7$
39. $x^4 + 3x^3 + 2x^2$
40. $x^7 - y^3x^4$
41. $x^2 - 2x + 1$
42. $10x^2 - 2x + 1$
43. Factoring as a difference of squares can be used to easily calculate the difference between two successive squares. For instance, $5^2 - 4^2 = (5+4)(5-4)$, and $5 - 4$ is 1. So a problem of squaring and subtraction reduces down to just addition.
    a. Compute $101^2 - 100^2$ by factoring as a difference of squares and then simplifying.
    b. Compute $85^2 - 84^2$ by factoring as a difference of squares and then simplifying.

Chapter 5 Review

5-A

1. Add $(x^3+3x^2+1)+(2x^3-x^2+1)$
2. Give the degree of the polynomial $2x^2+3y+1$.
3. Multiply and simplify $(2x+1)(x-4)$.
4. Multiply out and simplify $(x+2y)^2$.
5. Simplify $\frac{3x^2-7x}{x}$, assuming $x \neq 0$.
6. Divide, and indicate the quotient and remainder:

$$\frac{3x^2+10x-5}{x-1}$$

7. Divide, and indicate the quotient and remainder:

$$\frac{3x^3+2x^2-5x+1}{x^2+x+1}$$

8. Factor out the GCF: $28x-7x^2$.
9. Factor by grouping: $3x^2+xy+6x+2y$.
10. Factor the trinomial: x^2-4x-5.
11. Factor the difference of squares: x^2-25.
12. Factor the sum of cubes: x^3+1.
13. Factor completely: x^3+x^2-6x.

5-B

1. Simplify $(x^3+3x^2-7x-5)-(3x^3-2x^2-10x+1)$.
2. Give the degree of the polynomial $2x^2y^2+3xy^2+xy^3$.
3. Simplify $(2x^2+x+1)(x-1)$.
4. Multiply out and simplify $(3x-4y)^2$.

5. Simplify

$$\frac{x^3 + x^2 + 10x}{2x}.$$

6. Divide, and indicate the quotient and remainder:

$$\frac{7x^3 - x^2 + 10}{x + 3}$$

7. Divide, and indicate the quotient and remainder:

$$\frac{2x^4 - 5x^3 + 10x^2 + 7x - 1}{x^2 - 5x + 2}$$

Factor the following completely.

8. $100y^2 - 81x^2$
9. $3x^2 + 10x$
10. $3x^3 + xy - 12x^2 - 4y$
11. $3x^3 + 7x^2 + 2x$
12. $27x^3 + 125$
13. $6x^2 - x - 2$

CHAPTER

6

Rational Expressions and Equations

In this chapter we will consider rational expressions. Much like rational numbers are the quotients of integers, rational functions are the quotients of polynomials. It is important to know how to add, subtract, multiply, and divide these expressions, as well as how to solve rational equations and work with their applications. Anytime we are dealing with a proportion or a relative amount, we need to use fractions. So fractions with variable expressions in them are very useful—if we know how to use them correctly.

6.1: Multiplying and Dividing Rational Expressions

Multiplication

When multiplying rational expressions, make sure to factor them first. This makes it easy to see whether there will be any cancellation. This works the same when we multiply fractions that are just numbers. If we're multiplying

$$\frac{2}{4}\cdot\frac{4}{6},$$

we cancel the 4's and can factor the 6 as $2\cdot 3$ to get

$$\frac{2}{1}\cdot\frac{1}{2\cdot 3}.$$

From there, we can easily see that the 2's cancel and the answer is $\frac{1}{3}$. This same procedure will be used with variables.

Example: Simplify

$$\frac{x}{x+2}\cdot\frac{x^2-4}{x^2-x}.$$

Solution: Factor the difference of squares, and take an x out of the $x^2 - x$. Make liberal use of parentheses. The result is

$$\frac{x}{x+2}\cdot\frac{x^2-4}{x^2-x}=\frac{x}{(x+2)}\cdot\frac{(x+2)(x-2)}{x(x-1)}$$
$$=\frac{x-2}{x-1}.$$

Example: Simplify

$$\frac{4x^2-1}{x+1}\cdot\frac{x^2-4x-5}{2x^2-9x-5}.$$

Solution: First, we factor $4x^2 - 1$ as a difference of squares. Next, factoring $x^2 - 4x - 5$ as a trinomial, we have $(x-5)(x+1)$. Lastly, to factor $2x^2 - 9x - 5$ we start with $2x$ and x in our binomials; then our only choices are 1 and 5:

$$2x^2-9x-5=(2x+1)(x-5).$$

Now we write our factors out in preparation for canceling:

$$\frac{4x^2-1}{x+1}\cdot\frac{x^2-4x-5}{2x^2-9x-5}=\frac{(2x+1)(2x-1)}{x+1}\cdot\frac{(x-5)(x+1)}{(2x+1)(x-5)}$$
$$=2x-1.$$

Note that when we cancel everything in the denominator, we are left with $\frac{2x-1}{1}$, which equals $2x - 1$.

Division

Dividing by something is the same as multiplying by the reciprocal. So if we see a division symbol, the first step is to rewrite the second fraction flipped.

Example: Simplify

$$\frac{x^2-1}{x+2},\frac{x-1}{x^2-4}.$$

Solution: Be careful not to start factoring *until* we flip the sign and make it multiplication. So immediately change the equation to

$$\frac{x^2-1}{x+2}\cdot\frac{x^2-4}{x-1}.$$

Now, factoring gives us

$$\frac{(x+1)(x-1)}{x+2}\cdot\frac{(x+2)(x-2)}{x-1},$$

and canceling gives

$$(x+1)(x-2).$$

We should leave this as the answer. Since it is already factored, it's more simplified than if we multiplied it out.

Complex fractions have the form

$$\frac{\frac{a}{b}}{c}.$$

Always think of dividing by a number as being the same as multiplying by the reciprocal. In other words,

$$\frac{\frac{a}{b}}{c}=\frac{a}{b}\div c=\frac{a}{b}\cdot\frac{1}{c}$$

Note that we cannot interpret $a \div b \div c$ in terms of $a \div (b \div c)$. We must do division from left to right by the order of operations. In terms of the properties of real numbers, division is not associative, so we must take a *fraction* and *then* divide it by something else.

Example: Simplify

$$\frac{\frac{3x}{x+3}}{3x}$$

Solution: We must simplify this as $\frac{3x}{x+3}$ divided by $3x$. Dividing something by $3x$ is the same as multiplying it by $\frac{1}{3x}$. So we have

$$\begin{aligned}\frac{\frac{3x}{x+3}}{3x} &= \frac{3x}{x+3} \cdot \frac{1}{3x} \\ &= \frac{1}{x+3}.\end{aligned}$$

Now, if we have

$$\frac{\frac{a}{b}}{\frac{c}{d}},$$

this must be interpreted as $\frac{a}{b} \div \frac{c}{d}$, which is handled similarly to the example above (that is, we multiply by the reciprocal).

Example: Simplify

$$\frac{\frac{x^2+5x}{x^2-x}}{\frac{x^2+11x+30}{x+6}}.$$

Solution: This is one fraction divided by another. We take the first fraction, $\frac{x^2+5x}{x^2-x}$, and divide it by the second fraction, $\frac{x^2+11x+30}{x+6}$, to get this:

$$\begin{aligned}\frac{x^2+5x}{x^2-x}\div\frac{x^2+11x+30}{x+6}&=\frac{x^2+5x}{x^2-x}\cdot\frac{x+6}{x^2+11x+30}\\&=\frac{x(x+5)}{x(x-1)}\cdot\frac{x+6}{(x+5)(x+6)}\\&=\frac{1}{x-1}.\end{aligned}$$

Section 6.1 Exercises

Simplify the following multiplications, assuming that all denominators are nonzero.

1. $\frac{x-2}{x}\cdot\frac{x^2+4x}{x-2}$
2. $\frac{1}{x}\cdot\frac{x^2-4x}{x-4}$
3. $\frac{x^2-1}{x^2+x}\cdot\frac{x}{x^2-x}$
4. $\frac{x^2-1}{x+5}\cdot\frac{x^2-x-30}{x+1}$
5. $\frac{x^2-x-2}{x^2+x-6}\cdot\frac{x^2+7x+12}{x^2-6x-7}$
6. $\frac{x^2-x-30}{x^2-5x-6}\cdot\frac{x^2-x-2}{x^2+x-6}$
7. $\frac{x^2-x-12}{x^2+x-12}\cdot\frac{x^2+3x-4}{x^2+5x+6}$
8. $\frac{x^3-x}{x^2-x-2}\cdot\frac{x^3+x}{x^2-3x+2}$

9. $\frac{2x^2-x-1}{2x^2+3x-2}\cdot\frac{2x^2+5x-3}{2x^2+x}$

10. $\frac{3x^2+7x+2}{2x^2+x-3}\cdot\frac{2x^2-5x-12}{x^2-3x-10}$

Simplify the following divisions, assuming that all denominators are nonzero.

11. $\frac{x^2-2x-3}{x^2-2x-8}\div\frac{x^2-3x}{x^2-4x}$

12. $\frac{2x^2-x-1}{3x^2+14x+15}\div\frac{x^2+x-2}{3x^2-10x-25}$

13. $\frac{x^2-4}{x^2-9}\div\frac{x^2+x-2}{x^2-4x-21}$

14. $\frac{4x^2-25}{3x^2-5x-2}\div\frac{2x^2+7x+5}{3x^2+16x+5}$

15. $\dfrac{\frac{1}{2}}{x}$

16. $\dfrac{\frac{1}{x}}{2}$

17. $\dfrac{\frac{2x}{x+1}}{x}$

18. $\dfrac{\frac{8x-8}{3x+1}}{x-1}$

19. $\dfrac{\frac{x}{2}}{3}$

20. $\dfrac{\frac{x}{4}}{x}$

21. $\dfrac{\dfrac{2x^2+5x+3}{x-4}}{2x+3}$

22. $\dfrac{\dfrac{x-1}{2x+1}}{x}$

23. $\dfrac{\dfrac{x^2-x}{2x+5}}{x^2+5x-6}$

24. $\dfrac{\dfrac{7}{x}}{\dfrac{3}{x}}$

25. $\dfrac{\dfrac{2}{x}}{\dfrac{6}{x+1}}$

26. $\dfrac{\dfrac{x^2}{3}}{\dfrac{x}{5}}$

27. $\dfrac{\dfrac{x-5}{x-2}}{\dfrac{x^2-25}{x^2-4}}$

28. $\dfrac{\dfrac{x^2-9}{4x^2-81}}{\dfrac{x+3}{2x-9}}$

29. $\dfrac{\dfrac{x^2-x-2}{x^2-2x-15}}{\dfrac{x^2-3x-10}{x^2-9x-10}}$

6.2: Adding and Subtracting Rational Expressions

Addition

In the previous chapter, we discussed the greatest common factor, or GCF. Now we'll consider another term, called the "least common multiple." We use the least common multiple for finding denominators when we're adding fractions. The least common multiple is the smallest number that is a multiple of all numbers considered. For example, the least common multiple between 2 and 3 is 6, so when we add $\frac{1}{2}+\frac{1}{3}$ we know that the least common denominator is 6.

The same situation occurs when we're dealing with variable expressions. If we want to add $\frac{1}{x+2}$ and $\frac{1}{x+3}$, we must remember that x can be any number (except −2 and −3, which would give us division by zero). So the factors of the denominators are the numbers $(x+2)$ and $(x+3)$, and the least common denominator is $(x+2)(x+3)$. We then proceed the same way as before, multiplying each fraction by what it's missing from the least common denominator (LCD) and then adding.

Example: Simplify $\frac{1}{x+2}+\frac{1}{x+3}$.

Solution: We need a factor of $(x+2)$ and $(x+3)$ in our denominator. So the LCD is

$$(x+2)(x+3),$$

and we have

$$\begin{aligned}\frac{1}{x+2}+\frac{1}{x+3} &= \frac{(x+3)}{(x+2)(x+3)}+\frac{(x+2)}{(x+3)(x+2)}\\ &= \frac{x+3+x+2}{(x+2)(x+3)}\\ &= \frac{2x+5}{(x+2)(x+3)}.\end{aligned}$$

Example: Simplify $\frac{2}{x+5} - \frac{3}{x}$.

Solution: We have denominators of x and $x + 5$. Note that x is a number, and $x + 5$ is a number 5 units away from x, so these are different numbers. Our LCD must be the product of these two factors $x(x + 5)$:

$$\frac{2}{x+5} \cdot \frac{x}{x} - \frac{3}{x} \cdot \frac{(x+5)}{(x+5)}$$
$$= \frac{2x}{x(x+5)} - \frac{3x+15}{x(x+5)}.$$

Subtraction

Now, when we subtract the fraction in the above example, we bring the negative to the numerator and distribute it:

$$\frac{2x - 3x - 15}{x(x+5)}$$
$$= \frac{-x - 15}{x(x+5)}.$$

This cannot be simplified any further.

Example: Simplify $\frac{1}{x+4} + \frac{1}{x^2+2x-8}$.

Solution: Factoring the denominator, we have $(x + 4)(x - 2)$. So we are simplifying

$$\frac{1}{x+4} + \frac{1}{(x+4)(x-2)}.$$

Now, note that there is a factor of $x + 4$ in the first denominator, so that must be part of the LCD. Since there is a factor of $x + 4$ in the second denominator, we don't have to add another copy of it to the LCD. But there is also a factor of $x-2$ in the second denominator, so we must have $x-2$ in the LCD. The LCD is $(x+4)(x-2)$, so we have

$$\frac{1}{(x+4)}\cdot\frac{x-2}{(x-2)}+\frac{1}{(x+4)(x-2)}$$
$$\frac{x-2+1}{(x+4)(x-2)}$$
$$\frac{x-1}{(x+4)(x-2)}.$$

Again, we leave everything factored because it is easier to operate with that way.

Example: Simplify $\frac{1}{x^2-4}-\frac{3}{x^2-x-6}$.

Solution: Factoring the difference of squares gives $(x+2)(x-2)$. Factoring the trinomial gives $x^2-x-6=(x-3)(x+2)$. When we compare these, the first fraction tells us we need $(x+2)(x-2)$, and the second fraction tells us we need an $(x-3)$. Since we already have $(x+2)$ in our LCD, we don't need another one. So the LCD is

$$(x+2)(x-2)(x-3).$$

Simplifying, we have

$$\begin{aligned}\frac{1}{x^2-4}-\frac{3}{x^2-x-6}&=\frac{1}{(x+2)(x-2)}-\frac{3}{(x-3)(x+2)}\\&=\frac{1}{(x+2)(x-2)}\cdot\frac{x-3}{x-3}-\frac{3}{(x-3)(x+2)}\cdot\frac{x-2}{x-2}\\&=\frac{x-3}{(x+2)(x-2)(x-3)}-\frac{3(x-2)}{(x+2)(x-2)(x-3)}\\&=\frac{x-3-3(x-2)}{(x+2)(x-2)(x-3)}\\&=\frac{x-3-3x+6}{(x+2)(x-2)(x-3)}\\&=\frac{-2x+3}{(x+2)(x-2)(x-3)}.\end{aligned}$$

Recall the discussion in Section 1.2 where we said we can put a negative anywhere we like in the fraction. Consider the following example.

Example: Simplify $\frac{3}{x-2}-\frac{x}{2-x}$.

Solution: Note that we can factor a -1 out of $x-2$ to get $2-x=-(x-2)$. So really we need only one factor in the denominator, $x-2$. We have

$$\begin{aligned}\frac{3}{x-2}-\frac{x}{2-x}&=\frac{3}{x-2}-\frac{x}{-(x-2)}\\&=\frac{3}{x-2}+\frac{x}{x-2}\\&=\frac{3+x}{x-2}.\end{aligned}$$

Note that the negative in the denominator canceled with the subtraction in the first line above.

Example: What is the LCD needed to simplify the following?

$$\frac{x}{x^2-4}+\frac{1}{x^2+4x+4}+\frac{1}{x-2}$$

Solution: Factor the denominators:

$$\frac{x}{(x+2)(x-2)}+\frac{1}{(x+2)(x+2)}+\frac{1}{x-2}.$$

The first fraction tells us we need $(x+2)(x-2)$. The second fraction tells us we need two copies of $(x+2)$. So we multiply another factor of $(x+2)$ to get an LCD of

$$(x+2)(x+2)(x-2).$$

The third fraction tells us we need a factor of $x-2$, but we already have one and we do not need to add another. So our LCD is the same as above.

Section 6.2 Exercises

Simplify the following expressions.

1. $\frac{1}{2x}+\frac{1}{3x}$

2. $\frac{1}{x}+\frac{2}{x^2}$

3. $\frac{1}{2x+1}+\frac{1}{3x+1}$

4. $\frac{x}{x+3}-\frac{2}{x}$

5. $\frac{3}{x+2}+\frac{2}{x}$

6. $\frac{3}{x+1}+\frac{5}{x}$

7. $\frac{2}{x^2-x}+\frac{1}{x}$

8. $\frac{1}{x}-\frac{5}{x+1}$

9. $\frac{10}{x+2}-\frac{3}{x+4}$

10. $\frac{5}{x-8}+\frac{2}{x-5}$

11. $\frac{4}{2-x}+\frac{3}{x-2}$

12. $\frac{7}{x}+\frac{10}{-x}$

13. $\frac{4}{x-5}-\frac{3}{x-4}$

14. $\frac{2}{1-x}-\frac{3}{2-x}$

15. $\frac{x}{x^2+5x+6}-\frac{4}{x^2-4}$

16. $\frac{3}{x-4}-\frac{x}{2x-8}$

17. $\frac{3}{x^2+3x}-\frac{4}{x^2-9}$

18. $\frac{x-7}{(x+3)(x-2)}+\frac{4}{(x-2)(x+5)}$

19. $\frac{2x+10}{x^2+2x-3}-\frac{x}{x^2+7x+12}$

20. $\frac{2}{(x+7)(x-5)}-\frac{x}{(x+2)(x-5)}$

21. $\frac{1}{x^2-9}-\frac{3}{x^2-16}$

22. $\frac{1}{4x^2-25}-\frac{4}{9x^2-49}$

23. $\frac{1}{x}+\frac{2}{x^2}-\frac{3}{x^3}$

24. $\frac{1}{x+1}-\frac{3}{x^2-1}-\frac{4}{x-1}$

25. $\frac{2}{x} + \frac{3}{x^2 - 4x} - \frac{4}{x - 4}$

26. $\frac{1}{x^2 + 5x + 6} + \frac{1}{x^2 - x - 12} - \frac{3}{x^2 + 3x - 28}$

6.3: Rational Equations

In this section, we go from expressions to equations again. This distinction is important. In the *expression* $2x + 2$, we can factor out a 2 to get $2(x + 1)$. When we're working with the equation $2x + 2 = 0$, we can simply divide by 2 on both sides to get $\frac{2x+2}{2} = \frac{0}{2}$, or $x + 1 = 0$. We must be careful to know when we can just divide nonzero constants away, and when we have to keep the factored expression as it is.

When solving a rational equation, we are asking the question "For what x does this equation hold?" So it is important to know when an x will not work. We must be mindful of division by zero. So if we are solving the equation

$$\frac{1}{x} + \frac{3}{x + 1} = 0,$$

we immediately identify that x cannot be zero and $x + 1$ cannot be zero.

We can handle all this by checking our work. We've been somewhat lax about doing this, but as you'll see in this section, one of the formal steps for solving rational equations is to *check the answer*.

Here are the steps for solving a rational equation:

1. Find the LCD.
2. Multiply both sides by the LCD to clear the fractions.
3. Solve the resulting equation.
4. Check the answers in the original equation to make sure they work.

Example: Solve

$$\frac{1}{x}+\frac{3}{x+1}=0.$$

Solution: The LCD is the product of the denominators $x(x+1)$, so multiply both sides of the equation by $x(x+1)$ to clear the fractions:

$$x(x+1)\left[\frac{1}{x}+\frac{3}{x+1}\right]=x(x+1)[0].$$

Now distribute the $x(x+1)$ to each fraction on the left:

$$\begin{aligned}x(x+1)\cdot\frac{1}{x}+x(x+1)\cdot\frac{3}{x+1}&=0\\ x+1+x\cdot 3&=0\\ x+1+3x&=0\\ 4x+1&=0\\ 4x&=-1\\ x&=-\frac{1}{4}.\end{aligned}$$

To check that this actually works, plug it back in the original equation:

$$\begin{aligned}\frac{1}{-\frac{1}{4}}+\frac{3}{-\frac{1}{4}+1}&=-4+\frac{3}{\frac{3}{4}}\\ &=-4+3\cdot\frac{4}{3}\\ &=-4+4\\ &=0.\end{aligned}$$

So $x=-\frac{1}{4}$ is the solution to this equation.

Example: Solve

$$6+\frac{3x}{x+2}=-\frac{6}{x+2}.$$

Solution: The 6 in this equation has a denominator of 1, and the other fractions have denominators of $x+2$. So we multiply by $(x+2)$ on both sides to clear the fractions:

$$\begin{aligned}(x+2)\left[6+\frac{3x}{x+2}\right]&=(x+2)\left[-\frac{6}{x+2}\right]\\(x+2)\cdot 6+3x&=-6\\6x+12+3x&=-6\\9x+12&=-6\\9x&=-18\\x&=-2.\end{aligned}$$

Let's check this in the original equation. We have

$$6+\frac{3(-2)}{-2+2}=-\frac{6}{-2+2}$$

This gives us division by zero, which is not allowed. So $x=-2$ is not a solution. Since this was our only candidate to solve the equation, this equation has <u>no solutions.</u>

Example: Solve

$$\frac{3}{2x}+\frac{5}{3x}=\frac{3}{x-10}.$$

Solution: We need a factor of 2, 3, x, and $(x-10)$ in our LCD to clear the fractions. So we'll multiply both sides by $6x(x-10)$:

$$\begin{aligned} 6x(x-10)\left[\frac{3}{2x}+\frac{5}{3x}\right] &= 6x(x-10)\left[\frac{3}{x-10}\right] \\ 6x(x-10)\cdot\frac{3}{2x}+6x(x-10)\cdot\frac{5}{3x} &= 6x(3) \\ 3(x-10)\cdot 3+2(x-10)\cdot 5 &= 18x \\ 9(x-10)+10(x-10) &= 18x \\ 9x-90+10x-100 &= 18x \\ 19x-190 &= 18x \\ x-190 &= 0 \\ x &= 190. \end{aligned}$$

Let's check this in the original equation:

$$\begin{aligned} \frac{3}{2(190)}+\frac{5}{3(190)} &=?\frac{3}{190-10} \\ \frac{3}{380}+\frac{5}{570} &=?\frac{3}{180}. \end{aligned}$$

Using a calculator, we get

$$\frac{3}{380}+\frac{5}{570}=\frac{1}{60},$$

which is what $\frac{3}{180}$ simplifies to. So $x=190$ is our solution.

Example: Solve

$$\frac{2}{t+1}+\frac{8}{t-5}=\frac{10t+2}{t^2-4t+5}.$$

Solution: The quadratic expression $t^2 - 4t + 5$ can be factored into $(t+1)(t-5)$. So the LCD we'll multiply will be $(t+1)(t-5)$. This will clear the fractions. We have

$$(t+1)(t-5)\left[\frac{2}{t+1}+\frac{8}{t-5}\right]=(t+1)(t-5)\left[\frac{10t+2}{t^2-4t+5}\right]$$

$$(t+1)(t-5)\cdot\frac{2}{t+1}+(t+1)(t-5)\cdot\frac{8}{t-5}=10t+2$$

$$2(t-5)+8(t+1)=10t+2$$

$$2t-10+8t+8=10t+2$$

$$10t-2=10t+2$$

$$-2=2.$$

What happened here? We are asking for values of t that solve this equation, but any value of t we use will lead to a contradiction like 2 being equal to –2. There are no solutions to this problem.

Section 6.3 Exercises

Solve the following equations. Remember to check your answers.

1. $\frac{1}{x}+2=\frac{3}{x+1}$

2. $\frac{3}{x}=\frac{4}{x+1}$

3. $\frac{2}{x}-\frac{3}{x+1}=0$

4. $\frac{1}{3x+4}=\frac{1}{x-5}$

5. $2+\frac{1}{3x+5}=\frac{7}{6x+10}$

6. $\frac{3}{x}-\frac{4}{2x}=\frac{1}{3}$

7. $\frac{1}{2x}+\frac{3}{4x}=\frac{1}{3}$

8. $8-\frac{5}{x}+\frac{4}{2x}=0$

9. $\frac{7}{3x}+\frac{1}{2x}-\frac{5}{12x}=0$

10. $\frac{3}{x^2-3x-4}-\frac{5}{x-4}=-\frac{5}{x+1}$

11. $\frac{2}{x+2}-\frac{3}{x-2}=\frac{18}{x^2-4}$

12. $\frac{8}{x^2-2x-3}+\frac{2}{x+1}=\frac{2}{x-3}$

13. $\frac{3}{2x-3}-\frac{4}{2x^2+x-6}=-\frac{10}{x+2}$

14. $\frac{8}{x}=\frac{7}{x-1}$

15. $\frac{3}{x+2}-\frac{4}{x-1}=0$

16. $\frac{3}{x^2-x}=\frac{7}{x^2+4x-5}$

17. $\frac{2}{x^2-3x-4}=\frac{10}{x^2-1}$

18. $\frac{6}{x}+\frac{5}{2x}-\frac{2}{3x}+\frac{8}{12x}-1=0$

6.4: Applications and Proportions

In this section we will look at some of the applications to rational expressions. First we will do another round of solving formulas for variables, although this time they may be in the denominator of a fraction (as opposed to Section 2.2). Then we will look at some proportion problems.

Formulas

When we're solving for a specific variable in a formula that has several variables, we'll think of the other constants as numbers and we'll isolate the one we need to solve for.

Example: Solve the slope formula $m = \frac{y_2 - y_1}{x_2 - x_1}$ for x_2, assuming nonzero denominators.

Solution: The denominator of m is 1, and the denominator of the other fraction is $x_2 - x_1$. So this is the LCD that we multiply by on both sides:

$$(x_2 - x_1)m = (x_2 - x_1) \cdot \frac{y_2 - y_1}{x_2 - x_1}$$

$$mx_2 - mx_1 = y_2 - y_1$$

$$mx_2 = y_2 - y_1 + mx_1$$

$$x_2 = \frac{y_2 - y_1 + mx_1}{m}.$$

The instructions of this example say to assume nonzero denominators, so that we can ignore the case where the slope might be zero—because division by zero is undefined.

Example: Solve the following equation for a:

$$\frac{t}{a+b} + \frac{c}{b} = t.$$

Solution: The LCD here is $b(a+b)$. Multiplying this on both sides, we have

$$\begin{aligned} b(a+b)\left[\frac{t}{a+b}+\frac{c}{b}\right] &= b(a+b)\cdot t \\ b(a+b)\cdot\frac{t}{a+b}+b(a+b)\cdot\frac{c}{b} &= bt(a+b) \\ bt+c(a+b) &= bta+b^2t \\ bt+ca+cb &= bta+b^2t \\ bt+ca+cb-bta &= b^2t \\ ca-bta &= b^2t-cb-bt. \end{aligned}$$

We have moved everything with an a on it to one side, and everything without an a to the other side. Now we can factor a out from the left-hand side, which will allow us to isolate the variable:

$$\begin{aligned} a(c-bt) &= b^2-cb-bt \\ a &= \frac{b^2-cb-bt}{c-bt}. \end{aligned}$$

Note that this cannot be simplified, although we could factor b out in the numerator.

Word Problems

Sometimes, proportions

$$\frac{a}{b}=\frac{c}{d}$$

will be rational equations if there are variables in the denominator. We can clear the fractions by multiplying by the LCD, but this is equivalent to what is sometimes called cross-multiplying:

$$\frac{a}{b}=\frac{c}{d} \quad \text{is equivalent to} \quad ad=bc.$$

To see this, take $\frac{a}{b}=\frac{c}{d}$ and multiply both sides by bd:

$$bd\cdot\frac{a}{b}=bd\cdot\frac{c}{d}$$
$$ad=bc.$$

Cross-multiplying is helpful when we're dealing with proportion problems or conversion problems. Let's look at some examples.

Example: On a map of Scotland, 2 inches is equivalent to 75 miles. If Inverness and Edinburgh are 4.13 inches apart on the map, what is the distance between the cities of Inverness and Edinburgh in miles?

Solution: The conversion is

$$\frac{2\text{ in}}{75\text{ mi}}.$$

So if there are 4.13 inches between the two cities on the map, we let x be the distance between them:

$$\frac{2\text{ in}}{75\text{ mi}}=\frac{4.13\text{ in}}{x\text{ mi}}$$

We can cross-multiply to get

$$2x=75\cdot 4.13$$
$$2x=309.75$$
$$x=\frac{309.75}{2}$$
$$x=154.875,$$

which tells us that Edinburgh and Inverness are about 155 miles apart.

Example: A streetlight is 30 feet tall, and a woman 6 feet tall is walking away from the light. How long is the shadow she casts on a nearby wall when she is 10 feet away from the wall?

Solution: Consider the setup of the problem as depicted in Figure 6-1.

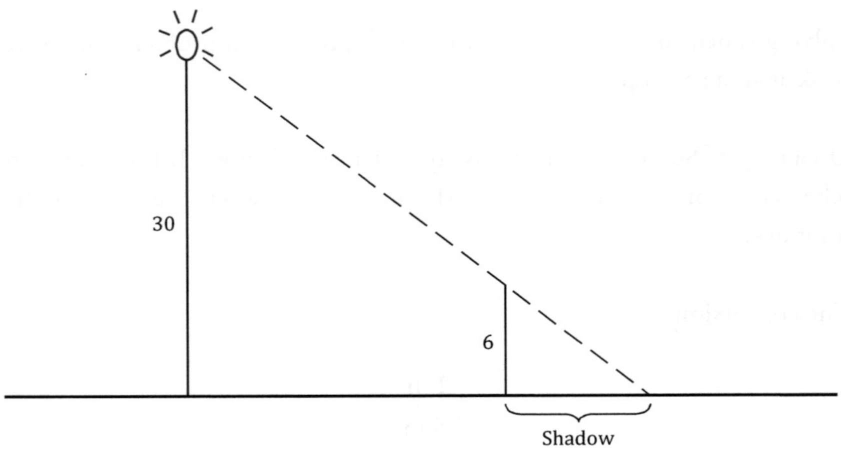

Fig. 6-1

If the length of the woman's shadow is x, and she is 10 feet away from the wall, we can use similar triangles to get the proportions depicted in Figure 6-2. Remember we take the ratio of the smaller triangle's sides to the larger triangle's sides.

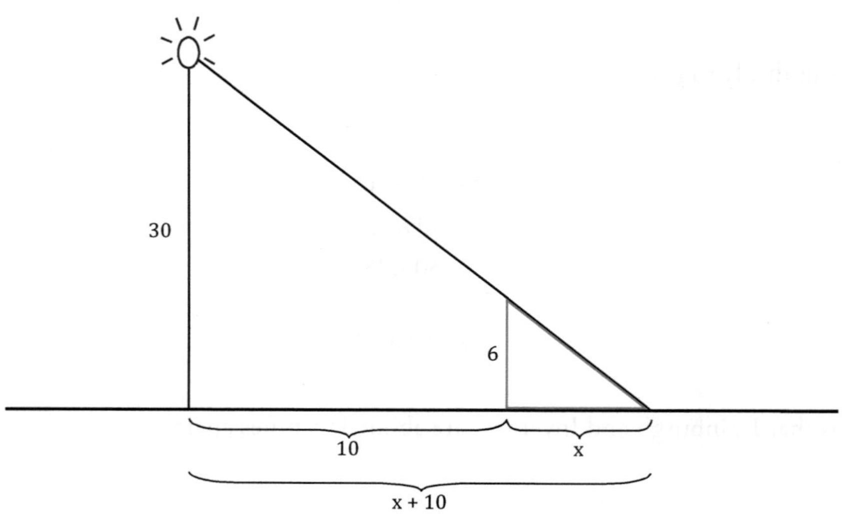

Fig. 6-2

$$\frac{6}{30} = \frac{x}{x+10}$$

We can cross-multiply here for

$$\begin{aligned} 6(x+10) &= 30x \\ 6x+60 &= 30x \\ 60 &= 24x \\ \frac{60}{24} &= x \\ \frac{15}{6} &= x. \end{aligned}$$

So her shadow is two and one-half feet long.

The formula $d = rt$ has been very useful in this text so far. It can be extended beyond velocity-distance problems. For instance, any time we know a rate and a time, we can multiply them together to get an amount:

$$\textit{Amount done} = \textit{rate} \cdot \textit{time}$$

Example: If Mrs. Terry can paint a house in 4 hours, and Mrs. Sowers can paint a house in 6 hours, how long will it take them to paint a house if they work together?

Solution: If Mrs. Terry can paint a house in 4 hours, then her rate is $\frac{1}{4}$ of a house per hour:

$$\begin{aligned} \textit{amount done} &= \textit{rate} \cdot \textit{time} \\ 1\ \textit{house} &= (\textit{rate}) \cdot (4\ \textit{hours}). \end{aligned}$$

Similarly, Mrs. Sowers' rate is $\frac{1}{6}$ of a house per hour. If Mrs. Terry and Mrs. Sowers work together, then their rate together would be the sum of their individual rates:

$$\begin{aligned} r_T &= \frac{1}{4} + \frac{1}{6} \\ &= \frac{6}{24} + \frac{4}{24} \\ &= \frac{10}{24} \\ &= \frac{5}{12}. \end{aligned}$$

This means they can paint $\frac{5}{12}$ of a house in one hour:

$$\begin{aligned} 1\ \textit{house} &= (r_T) \cdot \textit{time} \\ 1 &= \frac{5}{12} \cdot t \\ \frac{12}{5} &= t. \end{aligned}$$

So it will take them $\frac{12}{5}$ of an hour, or 2.4 hours to complete one house if they're working together.

Section 6.4 Exercises

Assume all variables are positive.

1. Solve the equation $r(1-s) = 5t - s$ for s.

2. Solve the equation $\frac{1}{2}p(q-z) = \frac{3}{2}q\left(p - \frac{1}{3}z\right)$ for z.

3. Recall from Section 1.6.2 that total resistance R_T can be expressed in terms of the resistors R_1 and R_2 connected in parallel using the equation

$$R_T = \left(\frac{1}{R_1} + \frac{1}{R_2}\right)^{-1}.$$

An equivalent way of writing this equation is

$$R_1R_2 = R_TR_2 + R_1R_T.$$

Solve this equation for R_1.

4. Solve the formula $3r + er = 4t$ for r.
5. Solve the following formula for c:

$$B = \frac{k}{c}$$

6. Solve the following formula for t:

$$\frac{a}{t} + \frac{b}{2t} = 1$$

7. Solve the following formula for m:

$$F = \frac{g}{m+1}$$

8. The total resistance R_T for two resistors R_1 and R_2 connected in parallel is given by the formula

$$\frac{1}{R_T} = \frac{1}{R_1} + \frac{1}{R_2}.$$

Solve this formula for R_1.

9. Drake is walking away from a streetlight that is 45 feet tall. If Drake is 7 feet tall, how much shadow is Drake casting when he is 8 feet from the streetlight?
10. If Dr. Beatty can paint two houses in 5 hours, and Dr. Schnackenberg can paint three houses in 7 hours, how long will it take them to paint five houses if they work together?

11. Two players of the Penguins basketball team, Stevie and Charlie, are practicing lay-ups. If it takes Stevie 45 seconds to do two lay-ups, and Charlie is twice as fast, how long will it take Charlie to do nine lay-ups?
12. In music, if you take the tempo and divide it by 60, you get the number of beats per second. If Warren is conducting his band at a fast presto tempo of 160 bpm, then how many beats per second are there?
13. Steve was able to solve four Analysis problems in the time it took me to solve one. How many Analysis problems does Steve solve in the time it takes me to solve five?
14. It costs $1.88 for 50 grams of bamboo cotton knitting yarn. If Maria spent $22.56 on the yarn, how many 50-gram skeins of yarn did Maria buy?
15. Jeff and his cat Sneak are able to help 4 out of every 5 people they meet. In a week, Jeff and Sneak helped 300 people. How many people did they meet that week?
16. Where Leah works, there are 9 men for every 2 women. If there are a total of 220 employees, how many men and how many women work with Leah?
17. Kyle needs one and one-half cups of flour to bake every dozen muffins. How much flour will Kyle need if he has to bake 42 muffins?
18. Stuart is able to watch seven movies for every three TV shows he watches. If Stuart watches a total of fifty programs (movies or TV), then how many movies did he watch and how many TV shows did he watch?
19. Katie needs 11 purple bows for every 3 red bows for the baskets she makes. If Katie used 280 bows this week, how many purple bows and how many red bows did she use?
20. Daniel can build a house and a half in a day and a half. Mark can build two houses in two days. How long will it take them to build one house if they're working together?
21. John launches a boat from shore. He is able to reach his destination (8 miles away) three times as fast (due to the coffee he had in the morning) as it takes for his return journey. If John's return journey was 16 hours long, how long did it take John to get where he was going?
22. Emily lives in Washington, D.C., and her friend Laura lives in Orlando. They're looking at a map that shows 1.5 inches as equivalent to 80 miles. If D.C. and Orlando are 16 inches apart on the map, how far apart do Emily and Laura live from each other in actual miles?
23. Crushman crushes cars at a speed that is triple that of the machine at the incinerator. If Crushman crushes 28 cars in 6 hours, how many cars will the machine have crushed in that time?

CHAPTER 6 REVIEW

6-A

1. Multiply and simplify

$$\frac{x}{x^2-4}\cdot\frac{x+2}{x^2-x}.$$

2. Divide and simplify

$$\frac{x^2-9}{x^2-2x-15}\div\frac{x+6}{x-5}$$

3. Simplify the complex fraction

$$\frac{\frac{x}{3}}{2x}.$$

4. Simplify the complex fraction

$$\frac{\frac{2}{x}}{\frac{3}{x^2}}.$$

5. Find the common denominator and add:

$$\frac{2}{x+5}+\frac{4}{x-7}.$$

6. Find the common denominator and add:

$$\frac{2}{3x}+\frac{4}{5x}.$$

7. Solve $\frac{2}{x}+\frac{6}{2x-1}=0$.

8. Chris can paint a kitchen in 7 hours, and Robert can paint a kitchen in 9 hours. How long will it take them to paint a kitchen if they work together?

6-B

Simplify the following expressions.

1. $\frac{x^2-x-12}{2x^2-13x-7}\cdot\frac{2x^2-5x-3}{x^2-7x+12}$

2. $\frac{2x^2-25x+50}{3x^2+19x+6}\div\frac{x^2-4x-60}{3x^2-17x-6}$

3. $\dfrac{\frac{2x-3}{4}}{4x-6}$

4. $\dfrac{\frac{6x}{4x+8}}{\frac{2x^2}{7}}$

5. $\frac{2}{x-7}-\frac{x}{x-6}$

6. $\frac{1}{x}+\frac{3}{2x^2}-\frac{8}{5x}$

7. Solve

$$\frac{2}{x+1}-\frac{4}{x}=6$$

8. If Janet can mow the yard in 3 hours, and Carol can mow the yard in 4 hours, how long will it take them to mow the yard if they work together?

CHAPTER

7

Radicals

We're going to dive deep into simplifying radical expressions and then solving radical equations. We will finish up by zooming out on our view of numbers to include nonreal numbers.

7.1: Rational Exponents

In this section, we will begin our discussion of radicals by understanding rational exponents. (In the next section, we will see how rational exponents deal directly with radicals; but for now, we will be using only the exponent rules for natural numbers and applying them to rational numbers.)

For example, how can we make $x^{\frac{1}{2}}$ make sense based on the fact that x^2 can be thought of as multiplying two copies of x together? We will have that $x^{\frac{1}{2}}$ will be some number α. What does this mean for α in terms of x? It would makes sense that if we square $x^{\frac{1}{2}}$, we should get x. In terms of our exponent rules, recall that a power raised to a power means multiplication. This way,

$$\left(x^{\frac{1}{2}}\right)^2 = x^{\frac{1}{2}\cdot 2} = x^1 = x.$$

So if $x^{\frac{1}{2}} = \alpha$, then squaring both sides means $x = \alpha^2$.

In the next section we will relate the idea of taking $x^{\frac{1}{2}}$ to taking the square root of x.

But this is why $4^{\frac{1}{2}} = 2$:

$$\begin{aligned} 4^{\frac{1}{2}} &= (2^2)^{\frac{1}{2}} \\ &= 2^{2\cdot\frac{1}{2}} \\ &= 2^1 \\ &= 2. \end{aligned}$$

If we are talking about the cube root, this would be taking something to the one-third power.

So $8^{\frac{1}{3}} = 2$, because $2^3 = 8$.

Consider this definition: If n is a natural number, x is positive, and $x^n = a$, then

$$a^{\frac{1}{n}} = x.$$

This makes sense, because if $2^4 = 16$, we can take the fourth root of 16 to get 2. That is,

$$16^{\frac{1}{4}} = (2^4)^{\frac{1}{4}} = 2^{4\cdot\frac{1}{4}} = 2^1 = 2.$$

Example: Simplify $25^{\frac{1}{2}} + 64^{\frac{1}{2}}$.

Solution: Since $5^2 = 25$ and $8^2 = 64$, we have

$$(5^2)^{\frac{1}{2}} + (8^2)^{\frac{1}{2}} = 5^{2\cdot\frac{1}{2}} + 8^{2\cdot\frac{1}{2}} = 5^1 + 8^1 = 13.$$

When we have to take a fractional power where the numerator is not 1, we often use the following important notation. If possible, we try to take care of the denominator first, which means that we evaluate $x^{\frac{a}{b}}$ by doing:

$$x^{\frac{a}{b}} = \left(x^{\frac{1}{b}}\right)^a.$$

Example: Simplify $27^{\frac{4}{3}}$.

Solution: When we have an exponent that is a fraction, we must take care of the denominator's action first. So we want to find a number whose cube is 27. That number is 3. Then we take 3 to the fourth power:

$$\begin{aligned} 27^{\frac{4}{3}} &= \left[(27)^{\frac{1}{3}}\right]^4 \\ &= [3]^4 \\ &= 81. \end{aligned}$$

Example: Simplify $8^{\frac{2}{3}}$.

Solution: Note that $2^3 = 8$. Simplifying the root first gives us

$$8^{\frac{2}{3}} = \left[8^{\frac{1}{3}}\right]^2 = [2]^2 = 4.$$

The rules for exponents that we learned in Chapter 1 can be extended to rational exponents. Recall that for $a,b > 0$, these are the rules:

$$\begin{aligned} a^m a^n &= a^{m+n} \\ \frac{a^m}{a^n} &= a^{m-n} \\ (a^m)^n &= a^{mn} \\ \left(\frac{a}{b}\right)^m &= \frac{a^m}{b^m} \\ x^{-n} &= \frac{1}{x^n} \\ \left(\frac{a}{b}\right)^{-m} &= \frac{b^m}{a^m}. \end{aligned}$$

Now we are ready to handle variable expressions.

Example: Simplify $\left(8x^{\frac{2}{3}}\right)\cdot\left(9x^{\frac{3}{2}}\right)$.

Solution: Everything is multiplied together here, so we multiply the 8 and the 9. And when we multiply variables with the same base, we add the exponents as shown here:

$$\begin{aligned} \left(8x^{\frac{2}{3}}\right)\cdot\left(9x^{\frac{3}{2}}\right) &= 8\cdot 9\cdot x^{\frac{2}{3}}\cdot x^{\frac{3}{2}} \\ &= 72x^{\frac{2}{3}+\frac{3}{2}} \\ &= 72x^{\frac{4}{6}+\frac{9}{6}} \\ &= 72x^{\frac{13}{6}}. \end{aligned}$$

Example: Simplify

$$\frac{x^{\frac{2}{3}}}{x^{-\frac{1}{3}}}.$$

Solution: Since we are dividing, we subtract the exponents:

$$\begin{aligned}\frac{x^{\frac{2}{3}}}{x^{-\frac{1}{3}}} &= x^{\frac{2}{3}-\left(-\frac{1}{3}\right)}\\ &= x^{\frac{2}{3}+\frac{1}{3}}\\ &= x^{\frac{3}{3}}\\ &= x.\end{aligned}$$

Example: Simplify

$$\frac{14x^2(3x+4)^{-\frac{5}{2}}}{7x^{-3}(3x+4)^{\frac{1}{2}}}.$$

Solution: We handle the constants, the x's, and the $(3x + 4)$'s separately. Everything is multiplied together, so we just divide the constants and subtract the exponents:

$$\begin{aligned}&\frac{14}{7}x^{2-(-3)}(3x+4)^{-\frac{5}{2}-\frac{1}{2}}\\ &= 2x^{2+3}(3x+4)^{-\frac{6}{2}}\\ &= 2x^5(3x+4)^{-3}\\ &= \frac{2x^5}{(3x+4)^3}.\end{aligned}$$

Example: Simplify

$$\frac{1}{2}x^{\frac{2}{3}}\left(4x^{-\frac{1}{3}}+18x^{\frac{3}{4}}\right).$$

Solution: Distributing, we have

$$\begin{aligned}\frac{1}{2}x^{\frac{2}{3}}\left(4x^{-\frac{1}{3}}+18x^{\frac{3}{4}}\right)&=\frac{1}{2}x^{\frac{2}{3}}\left(4x^{-\frac{1}{3}}\right)+\frac{1}{2}x^{\frac{2}{3}}\left(18x^{\frac{3}{4}}\right)\\&=2x^{\frac{2}{3}-\frac{1}{3}}+9x^{\frac{2}{3}+\frac{3}{4}}\\&=2x^{\frac{1}{3}}+9x^{\frac{8}{12}+\frac{9}{12}}\\&=2x^{\frac{1}{3}}+9x^{\frac{17}{12}}.\end{aligned}$$

When we are dealing with rational exponents, being able to factor the expression will often be beneficial. For example, if factoring will make a sum of two terms into one term, then it will be worth it.

To do this, we factor out the GCF, which takes out as much as possible that is common to all terms. So between x^2 and x, the GCF is x. Another way to approach it is to take out the smallest exponent. So when we factor $x^2 - x$, we can take out an x. The same applies to rational exponents: *Take out the factor with the smallest exponent.*

Example: Factor

$$(x+3)^{\frac{1}{2}}+\frac{1}{2}x(x+3)^{-\frac{1}{2}}.$$

Solution: We have one term with $(x+3)^{\frac{1}{2}}$ and another term with $(x+3)^{-\frac{1}{2}}$. Remember that when we're factoring with rational exponents, the GCF is taking out as much as possible that is common to both. This means we take out only the smallest exponent, which is the $x+3$ raised to the $-\frac{1}{2}$ power:

$$(x+3)^{-\frac{1}{2}}[?+?]$$

Now, to determine what's left, when we factor something out we are dividing it. So in the first term we have $(x+3)^{\frac{1}{2}}$ and we're dividing out an $(x+3)^{-\frac{1}{2}}$. What's left is

$$\frac{(x+3)^{\frac{1}{2}}}{(x+3)^{-\frac{1}{2}}} = (x+3)^{\frac{1}{2}-\left(-\frac{1}{2}\right)} = (x+3)^1 = x+3.$$

Now we are ready to finish the problem:

$$\begin{aligned}(x+3)^{\frac{1}{2}} + \frac{1}{2}x(x+3)^{-\frac{1}{2}} &= (x+3)^{-\frac{1}{2}}\left[(x+3) + \frac{1}{2}x\right] \\ &= (x+3)^{-\frac{1}{2}}\left(x+3+\frac{1}{2}x\right) \\ &= (x+3)^{-\frac{1}{2}}\left(\frac{3}{2}x+3\right).\end{aligned}$$

Recall that expressions that are factored and are all multiplied together are more simplified than expressions with multiple terms. This is much better than adding the two terms we began with.

Example: Factor and simplify

$$-\frac{1}{2}(x+3)^{-\frac{3}{2}}\left(\frac{3}{2}x+3\right) + \frac{3}{2}(x+3)^{-\frac{1}{2}}.$$

Solution: There are two terms here, the first with a factor of $(x+3)^{-\frac{3}{2}}$ and the second with a factor of $(x+3)^{-\frac{1}{2}}$. The GCF between the two would be $(x+3)^{-\frac{3}{2}}$ (because $-\frac{3}{2}$ is less than $-\frac{1}{2}$). Note that both terms also have a $\frac{1}{2}$. So we factor out $\frac{1}{2}(x+3)^{-\frac{3}{2}}$ and get

$$\frac{1}{2}(x+3)^{-\frac{3}{2}}\left[-\left(\frac{3}{2}x+3\right)+\frac{\frac{3}{2}(x+3)^{-\frac{1}{2}}}{(x+3)^{-\frac{3}{2}}}\right]$$

$$=\frac{1}{2}(x+3)^{-\frac{3}{2}}\left[-\frac{3}{2}x-3+\frac{3}{2}(x+3)\right]$$

$$=\frac{1}{2}(x+3)^{-\frac{3}{2}}\left[-\frac{3}{2}x-3+\frac{3}{2}x+\frac{9}{2}\right]$$

$$=\frac{1}{2}(x+3)^{-\frac{3}{2}}\left[-\frac{6}{2}+\frac{9}{2}\right].$$

$$=\frac{1}{2}(x+3)^{-\frac{3}{2}}\left(\frac{3}{2}\right)$$

$$=\frac{3}{4}(x+3)^{-\frac{3}{2}}$$

$$=\frac{3}{4(x+3)^{\frac{3}{2}}}.$$

Section 7.1 Exercises

1. Simplify $36^{\frac{1}{2}}$.
2. Simplify $81^{\frac{1}{4}}$.
3. Simplify $49^{\frac{1}{2}}$.
4. Simplify $-25^{\frac{1}{2}}$.
5. Simplify $(-8)^{\frac{1}{3}}$.

6. Simplify $(-27)^{\frac{2}{3}}$.

7. Simplify $27^{\frac{2}{3}}$.

8. Simplify $16^{\frac{1}{4}} - 16^{\frac{1}{2}}$.

9. Simplify $64^{\frac{2}{3}}$.

10. a. Simplify $64^{\frac{1}{6}}$.

    b. Simplify $64^{\frac{5}{6}}$.

11. Simplify $25^{-\frac{1}{2}}$.

12. Simplify $100^{-\frac{3}{2}}$.

13. Simplify $\left(\frac{8}{27}\right)^{-\frac{1}{3}}$.

14. Simplify $\left(\frac{4}{9}\right)^{\frac{1}{2}}$.

15. Simplify $\left(\frac{4}{25}\right)^{-\frac{3}{2}}$.

16. Simplify

$$\frac{x^{\frac{8}{3}}}{x^{\frac{4}{3}}}.$$

17. Simplify

$$\frac{4x^{\frac{1}{4}}}{2x^{\frac{3}{4}}}.$$

18. Simplify

$$\frac{40x^{\frac{5}{2}}y^{\frac{7}{9}}}{80x^{\frac{3}{2}}y^{\frac{1}{9}}}.$$

19. Simplify

$$\frac{28x^{\frac{1}{2}}(x+4)^{\frac{2}{3}}}{2x^{-\frac{3}{2}}(x+4)^{\frac{1}{3}}}$$

20. Factor $x^{\frac{7}{2}}+x^{\frac{5}{2}}+x^{\frac{1}{2}}$.

21. Factor $2x^{-\frac{3}{2}}y^{\frac{1}{2}}+4x^{\frac{1}{2}}y^{-\frac{3}{2}}$.

22. Factor $15x^{\frac{5}{2}}+5x^{\frac{3}{2}}$.

23. Factor

$$(2x-1)^{\frac{1}{4}}+\frac{1}{2}x(2x-1)^{-\frac{3}{4}}.$$

24. Factor

$$4(3x+5)^{\frac{1}{3}}+12x(3x+5)^{-\frac{2}{3}}.$$

25. Factor

$$4x(x+2)^{-\frac{2}{3}}+8x^2(x+2)^{-\frac{5}{3}}+6x^3(x+2)^{-\frac{2}{3}}.$$

26. Factor

$$12(3x-2)^{\frac{1}{2}}-20x(3x-2)^{-\frac{1}{2}}.$$

7.2: Radical Expressions

In this section we make precise the relationship between fractional exponents and radicals.

The square root symbol is used to denote a number raised to the one-half power. So

$$\sqrt{4} = 4^{\frac{1}{2}}.$$

Both of these expressions are equal to 2.

For the cube root, we use the radical symbol $\sqrt{\ }$ with a different index. For example,

$$\sqrt[3]{x}$$

is the symbol for the "cube root of x." So $\sqrt[3]{8} = 2$, because $2^3 = 8$. In general,

$$\boxed{\sqrt[n]{x} = x^{\frac{1}{n}}}$$

Ex) $\sqrt[4]{16} = 2$ because $16^{\frac{1}{4}} = 2$.

Ex) $\sqrt[3]{27} = 3$ because $27^{\frac{1}{3}} = 3$.

We are now ready to make the definition formal:

$$x^{\frac{m}{n}} = \sqrt[n]{x^m}.$$

This is precisely why we say the denominator in an exponent is a "root,": because we are taking the nth root of a number. We should always try to do the root first if we can, and then raise that number to the mth power.

Recall that $(ab)^{\frac{1}{n}} = a^{\frac{1}{n}} b^{\frac{1}{n}}$, as long as a and b are positive. This is the property we are using when we simplify radicals, but we're using the radical notation:

$$\sqrt[n]{ab} = \sqrt[n]{a}\sqrt[n]{b},\ a,b > 0.$$

Example: Simplify $\sqrt{18}$.

Solution: Since $18 = 9 \cdot 2$, we have

$$\sqrt{18} = \sqrt{9 \cdot 2} = \sqrt{9} \cdot \sqrt{2} = 3\sqrt{2}.$$

When simplifying a cube root, we simplify in triplets.

Example: Simplify $\sqrt[3]{54}$.

Solution: Prime-factoring 54, we have $54 = 27 \cdot 2$. So we have

$$\begin{aligned} \sqrt[3]{54} &= \sqrt[3]{27}\sqrt[3]{2} \\ &= 3\sqrt[3]{2}. \end{aligned}$$

Example: Simplify $\sqrt{800}$.

Solution: First we factor 800, and take out as many squares as possible. The result:

$$\begin{aligned} \sqrt{800} &= \sqrt{8 \cdot 100} \\ &= \sqrt{8 \cdot 25 \cdot 4} \\ &= \sqrt{4 \cdot 2 \cdot 25 \cdot 4} \\ &= \sqrt{4}\sqrt{2}\sqrt{25}\sqrt{4} \\ &= 2\sqrt{2} \cdot 5 \cdot 2 \\ &= 20\sqrt{2}. \end{aligned}$$

Example: Simplify $\sqrt{x^3 y^5}$, assuming x and y are both nonnegative.

Solution: We can simplify $\sqrt{x^2}$ to be x (as long as x is nonnegative; see the discussion below about even roots). We will write x^3 as $x^2 \cdot x$, so that we can simplify the radical. Similarly, we can write y^5 as $y^4 \cdot y$, so the square root of y^4 will simplify to $y^{\frac{4}{2}} = y^2$. So we have

$$\begin{aligned}\sqrt{x^3 y^5} &= \sqrt{x^2 \cdot x \cdot y^4 \cdot y} \\ &= \sqrt{x^2} \cdot \sqrt{y^4}\sqrt{xy} \\ &= xy^2\sqrt{xy}.\end{aligned}$$

We will not leave exponents higher than 2 in a square root simplification. Likewise, we will not leave any exponents higher than 3 in a cube root.

Example: Simplify $\sqrt[3]{16x^{10}y^8}$.

Solution: We want to take out of the radical exponents that are multiples of 3. We are taking the cube root, which is like raising these expressions to the one-third power. We have

$$\begin{aligned}\sqrt[3]{16x^{10}y^8} &= \sqrt[3]{8 \cdot 2 \cdot x^9 \cdot x \cdot y^6 \cdot y^2} \\ &= \sqrt[3]{8} \cdot \sqrt[3]{x^9} \cdot \sqrt[3]{y^6} \cdot \sqrt[3]{2xy^2} \\ &= 2x^3y^2 \cdot \sqrt[3]{2xy^2}.\end{aligned}$$

Even Roots

By *even roots* we mean square roots or fourth roots, etc. Consider the expression

$$\sqrt{x^2}.$$

If $x = 6$, then $\sqrt{6^2} = \sqrt{36} = 6$.

If $x = -6$, then $\sqrt{(-6)^2} = \sqrt{36} = 6$.

This is interesting, because no matter what x is, if we do the operation $\sqrt{x^2}$, the result is always positive (or it's zero if $x = 0$). Another expression that always returns a nonnegative version of x is **absolute value**.

Indeed, if $x > 0$, then $\sqrt{x^2} = x$. But if $x < 0$, then $\sqrt{x^2} = -x$ (if x is negative, to make it positive we have to take its *opposite*). This is precisely the definition of *absolute value*, as in the following equation:

$$\sqrt{x^2} = |x|.$$

But this holds if we take any even root of an even power. For example, $\sqrt[4]{x^4} = |x|$.

Example: Simplify $\sqrt{x^2 y^6}$.

Solution: Since x and y could be either positive or negative, we have to remember the absolute value:

$$\begin{aligned}\sqrt{x^2 y^6} &= \sqrt{x^2}\sqrt{(y^2)^3} \\ &= |x||y|^3.\end{aligned}$$

Odd Roots

Odd roots are a lot less restrictive. We can take cube roots of negative numbers. If x is nonnegative or if x is negative,

$$\sqrt[3]{x^3} = x.$$

Section 7.2 Exercises

Simplify assuming variables represent positive real numbers.

1. $\sqrt{128}$
2. $\sqrt{75}$

3. $\sqrt{28}$

4. $\sqrt{150}$

5. $\sqrt[3]{81}$

6. $\sqrt[3]{24}$

7. $\sqrt[3]{216}$

8. $\sqrt[3]{-\frac{1}{8}}$

9. $\sqrt[3]{-\frac{27}{64}}$

10. $\frac{\sqrt{50}}{\sqrt{5}}$

11. $\sqrt[4]{32}$

12. $\frac{\sqrt{75}}{\sqrt{5}}$

13. Simplify by performing the root first: $\sqrt[5]{32^2}$

14. Simplify by performing the root first: $\sqrt{16^3}$

15. $\sqrt{x^4}$

16. $\sqrt{n^3}$

17. $\sqrt{n^8 y^9}$

18. $\sqrt{m^{13}}$

19. $\sqrt[3]{p^{13} q^{15}}$

20. $\sqrt[4]{16y^5}$

21. $\sqrt[3]{108x^7 y^6}$

22. $\sqrt[4]{z^{12}v^{19}}$

23. $\sqrt{49x^2 y^{10}}$

24. $\sqrt[3]{36x^{-6} y^{-9}}$

25. $\sqrt{100x^{-4}}$

26. $\sqrt[3]{-\frac{64}{x^{-3}}}$

Simplify exercises 27–30, assuming variables are real numbers.

27. $\sqrt{x^6 y^{10}}$

28. $\sqrt{49x^2 z^6}$

29. $\sqrt[3]{-x^3}$

30. $\sqrt[3]{-u^6 w^9}$

7.3: Adding and Subtracting Radicals

In this section we will add and subtract radical expressions. This is similar to things we have done before, because it's precisely adding like terms. For example, if we add $\sqrt{2}$ to $\sqrt{2}$, we have two copies of $\sqrt{2}$, which we write as $2\sqrt{2}$.

Another reason why it's easier to think about like terms is that we cannot simplify $\sqrt{2}+\sqrt{3}$. We need what's inside the radical to be the same. So consider $\sqrt{2}$ and $\sqrt{3}$ to be different terms.

Now, this can be deceiving, because it may not look like $\sqrt{2}+\sqrt{50}$ can be simplified. But if we prime-factor 50, we know that $50=25\cdot 2$. Taking the square root of 50 is identical to $\sqrt{25\cdot 2}$, which is $5\sqrt{2}$. The following example makes this clear.

Example: Simplify $\sqrt{2}+\sqrt{50}$.

Solution: We have

$$\begin{aligned}\sqrt{2}+\sqrt{50} &= \sqrt{2}+\sqrt{25\cdot 2}\\ &= \sqrt{2}+\sqrt{25}\sqrt{2}\\ &= \sqrt{2}+5\sqrt{2}\\ &= 6\sqrt{2}.\end{aligned}$$

It is helpful to think of $\sqrt{2}$ and $5\sqrt{2}$ as like terms, because we add their coefficients (1 and 5) together to get 6 square roots of 2.

Another way to think about this: If we factor out $\sqrt{2}$ from both expressions, then

$$\begin{aligned}\sqrt{2}+5\sqrt{2} &= \sqrt{2}(1+5)\\ &= \sqrt{2}\cdot 6\\ &= 6\sqrt{2}.\end{aligned}$$

Example: Simplify $\sqrt{98}+\sqrt{18}-\sqrt{200}$.

Solution: Start by dividing what's inside the radicals by 2:

$$\begin{aligned}\sqrt{98}+\sqrt{18}-\sqrt{200} &= \sqrt{49\cdot 2}+\sqrt{9\cdot 2}-\sqrt{100\cdot 2}\\ &= \sqrt{49}\sqrt{2}+\sqrt{9}\sqrt{2}-\sqrt{100}\sqrt{2}\\ &= 7\sqrt{2}+3\sqrt{2}-10\sqrt{2}\\ &= 10\sqrt{2}-10\sqrt{2}\\ &= 0.\end{aligned}$$

Example: Simplify $\sqrt[3]{-27}-\sqrt[3]{-8}$.

Solution: Since $(-3)^3=-27$ and $(-2)^3=-8$, we have

$$\begin{aligned}\sqrt[3]{-27}-\sqrt[3]{-8} &= -3-(-2)\\ &= -3+2\\ &= -1.\end{aligned}$$

Of course, we have to be mindful if the radicals do not match up.

Example: Simplify $\sqrt[3]{2}\cdot\sqrt[3]{16}+\sqrt[4]{27}\cdot\sqrt[4]{9}$.

Solution: Because the radicals being multiplied have the same index, they can be put under the same radical. Then we will look for roots to take out. We have

$$\begin{aligned}\sqrt[3]{2}\cdot\sqrt[3]{16}+\sqrt[4]{27}\cdot\sqrt[4]{9} &= \sqrt[3]{2\cdot 16}+\sqrt[4]{27\cdot 9}\\ &= \sqrt[3]{2\cdot 8\cdot 2}+\sqrt[4]{9\cdot 3\cdot 9}\\ &= \sqrt[3]{8\cdot 4}+\sqrt[4]{3\cdot 3\cdot 3\cdot 3\cdot 3}\\ &= \sqrt[3]{8}\cdot\sqrt[3]{4}+\sqrt[4]{3\cdot 3\cdot 3\cdot 3}\cdot\sqrt[4]{3}\\ &= 2\sqrt[3]{4}+3\sqrt[4]{3}.\end{aligned}$$

Example: Assuming $x>0$, simplify $\sqrt{6x}+\sqrt{24x}$.

Solution: We have

$$\begin{aligned}\sqrt{6x}+\sqrt{24x} &= \sqrt{6x}+\sqrt{4\cdot 6x}\\ &= \sqrt{6x}+2\sqrt{6x}\\ &= 3\sqrt{6x}.\end{aligned}$$

Recall that the area of a circle is $A=\pi r^2$. If a circle has area A, then the radius in terms of the area is

$$r=\sqrt{\frac{A}{\pi}}.$$

Example: If the area of a circle is 50 square meters, find its radius to the nearest tenth.

Solution: If the area is 50 square meters, then the radius is

$$\begin{aligned}r &= \sqrt{\frac{50}{\pi}}\\ &= \sqrt{15.915\ldots}\\ &= 3.989\ldots.\end{aligned}$$

So the radius rounded to the nearest tenth is 4.0.

Section 7.3 Exercises

Simplify, assuming all variables are nonnegative.

1. $\sqrt{8}+\sqrt{12}$
2. $\sqrt{28}+\sqrt{63}$
3. $\sqrt{10}+\sqrt{40}$
4. $\sqrt[3]{16}-\sqrt[3]{24}$
5. $\sqrt[3]{32}+\sqrt[3]{-4}$
6. $\sqrt{128}+\sqrt{72}-\sqrt{98}$
7. $\sqrt{24}-3\sqrt{54}+2\sqrt{125}$
8. $\sqrt[3]{24}-3\sqrt[3]{81}+\sqrt[3]{27}$
9. $\sqrt[3]{48}-\sqrt[3]{-48}$
10. $\sqrt{18}+\sqrt{6}+\sqrt{8}$
11. $\sqrt{8x}+\sqrt{200x}$
12. $\sqrt{48x}+\sqrt{75x}$
13. $\sqrt{700x^3}+x\sqrt{81x}$
14. $\sqrt{121x^2}-x$
15. $\sqrt{6x}+\sqrt{54x}-\sqrt{96x}$
16. $\sqrt[3]{54x}+\sqrt[3]{16x}$
17. $\sqrt[3]{81x^4}-2x\sqrt[3]{x}$

18. If the area of a circle is 100 square meters, find its radius to the nearest tenth.
19. If the area of a circle is 100π square meters, find the exact value of its radius.

7.4: Multiplying and Dividing Radicals

In this section we will deal with the multiplication and division of radicals. As in Chapters 5 and 6, we need to learn how the addition, multiplication, and division of these objects work so we will be able to solve their equations and do applications.

In Sections 7.2 and 7.3, we used the fact that

$$\sqrt[n]{ab} = \sqrt[n]{a}\sqrt[n]{b}, (a,b > 0).$$

In this section, to multiply to radicals, we will move from right to left in the above equation.

Ex) $\sqrt{2}\sqrt{3}$ can be simplified to one radical by bringing the 2 and 3 under one radical $\sqrt{2\cdot 3} = \sqrt{6}$.

Important: Remember to bring the numbers under one radical but not to multiply them together, so we can simplify.

Ex) If we are simplifying $\sqrt{6}\sqrt{10}$, we bring the numbers under the same "house": $\sqrt{6\cdot 10}$. But we'll resist the temptation to multiply and get 60, because we want to simplify this. So

$$\begin{aligned}\sqrt{6}\sqrt{10} &= \sqrt{6\cdot 10}\\ &= \sqrt{2\cdot 3\cdot 2\cdot 5}\\ &= \sqrt{4}\sqrt{3\cdot 5}\\ &= 2\sqrt{15}.\end{aligned}$$

Example: Simplify $(2+\sqrt{3})(2-\sqrt{3})$.

Solution: To simplify this, we have to FOIL it out:

$$\begin{aligned}(2+\sqrt{3})(2-\sqrt{3}) &= 4-\sqrt{3}\sqrt{3}+2\sqrt{3}-2\sqrt{3}\\ &= 4-3\\ &= 1.\end{aligned}$$

Note that we just simplified something of the form $(a+b)(a-b)$, so it makes sense that the cross terms cancel.

Example: Simplify $(\sqrt{2}+7\sqrt{5})(3\sqrt{2}+\sqrt{5})$.

Solution: FOIL this out for

$$\begin{aligned}(\sqrt{2}+7\sqrt{5})(3\sqrt{2}+\sqrt{5}) &= \sqrt{2}\cdot 3\sqrt{2}+\sqrt{2}\sqrt{5}+21\sqrt{5}\sqrt{2}+7\sqrt{5}\sqrt{5}\\ &= 3\sqrt{2\cdot 2}+\sqrt{10}+21\sqrt{10}+7\sqrt{5\cdot 5}\\ &= 3\cdot 2+22\sqrt{10}+7\cdot 5\\ &= 41+22\sqrt{10}.\end{aligned}$$

Example: Simplify $(4\sqrt{2}+3\sqrt{2x})(5\sqrt{2}-7\sqrt{2x})$.

Solution: We FOIL for

$$\begin{aligned}(4\sqrt{2}+3\sqrt{2x})(5\sqrt{2}-7\sqrt{2x}) &= 4\sqrt{2}\cdot 5\sqrt{2}-4\sqrt{2}\cdot 7\sqrt{2x}+3\sqrt{2x}\cdot 5\sqrt{2}-3\sqrt{2x}\cdot 7\sqrt{2x}\\ &= 20\sqrt{2\cdot 2}-28\sqrt{2\cdot 2\cdot x}+15\sqrt{2\cdot x\cdot 2}-21\sqrt{2\cdot x\cdot 2\cdot x}\\ &= 20\cdot 2-28\cdot 2\sqrt{x}+15\cdot 2\sqrt{x}-21\cdot 2\cdot x\\ &= 40-56\sqrt{x}+30\sqrt{x}-42x\\ &= 40-26\sqrt{x}-42x.\end{aligned}$$

Example: Simplify $\sqrt[3]{6x}(\sqrt[3]{36x^2}-\sqrt[3]{72x^5})$.

Solution: We distribute the $\sqrt[3]{6x}$ to each term inside the parentheses. So we have

$$\begin{aligned}\sqrt[3]{6x}(\sqrt[3]{36x^2}-\sqrt[3]{72x^5}) &= \sqrt[3]{6x}\cdot\sqrt[3]{36x^2}-\sqrt[3]{6x}\cdot\sqrt[3]{72x^5}\\ &= \sqrt[3]{6x\cdot 36x^2}-\sqrt[3]{6x\cdot 72x^5}\end{aligned}$$

Be careful not to multiply the constants together. We should factor $6\cdot 36$ and $6\cdot 72$ apart to see what cubes we can take out.

$$\begin{aligned}&= \sqrt[3]{6\cdot 6^2\cdot x^3}-\sqrt[3]{6\cdot 36\cdot 2\cdot x^6}\\ &= 6x-6x^2\sqrt[3]{2}.\end{aligned}$$

Example: Simplify by multiplying out $(\sqrt[3]{9v^2}+2\sqrt[3]{3v})^2$.

Solution: Remember that we can write this as the binomial times itself. So we have

$$\begin{aligned}(\sqrt[3]{9v^2}+2\sqrt[3]{3v})^2 &= (\sqrt[3]{9v^2}+2\sqrt[3]{3v})(\sqrt[3]{9v^2}+2\sqrt[3]{3v})\\ &= \sqrt[3]{9v^2}\cdot\sqrt[3]{9v^2}+\sqrt[3]{9v^2}\cdot 2\sqrt[3]{3v}+2\sqrt[3]{3v}\cdot\sqrt[3]{9v^2}+2\sqrt[3]{3v}\cdot 2\sqrt[3]{3v}\\ &= \sqrt[3]{9\cdot 9\cdot v^2\cdot v^2}+2\sqrt[3]{9\cdot 3\cdot v^2\cdot v}+2\sqrt[3]{3\cdot 9\cdot v\cdot v^2}+4\sqrt[3]{3\cdot 3\cdot v\cdot v}\\ &= 3v\sqrt[3]{3v}+6v+6v+4\sqrt[3]{9v^2}\\ &= 3v\sqrt[3]{3v}+12v+4\sqrt[3]{9v^2}.\end{aligned}$$

Dividing Radicals

The process of simplifying the division of radicals is sometimes called rationalization. Before there were calculators, rationalizing was done a lot using tables for square roots. From practice, many mathematicians would know that $\sqrt{2}$ is approximately 1.414. If their calculations led them to

$$\frac{1}{\sqrt{2}},$$

they would multiply the numerator and denominator by $\sqrt{2}$ as follows:

$$\begin{aligned}\frac{1}{\sqrt{2}} &= \frac{1}{\sqrt{2}} \cdot \frac{\sqrt{2}}{\sqrt{2}} \\ &= \frac{\sqrt{2}}{2}.\end{aligned}$$

They did this because it's much easier to take our approximation for $\sqrt{2}$ and divide it by 2 than it is to calculate $\frac{1}{\sqrt{2}}$.

This process hardly seems necessary now, but it is. Rationalizing helps us simplify radicals where otherwise we would just have confusing expressions.

What if we have something like $\frac{x}{\sqrt{x}}$? If we rationalize, we see that this is just $\sqrt{x}$.

Ex) If $x > 0$, then $\frac{x}{\sqrt{x}} = \frac{x}{\sqrt{x}} \cdot \frac{\sqrt{x}}{\sqrt{x}} = \frac{x\sqrt{x}}{x} = \sqrt{x}$.

Ex) $\frac{2}{\sqrt{2}}$ can be simplified to

$$\begin{aligned}\frac{2}{\sqrt{2}} &= \frac{2}{\sqrt{2}} \cdot \frac{\sqrt{2}}{\sqrt{2}} \\ &= \frac{2\sqrt{2}}{2} \\ &= \sqrt{2}.\end{aligned}$$

Ex) $\frac{1}{\sqrt{5}}$ can be rationalized to

$$\begin{aligned}\frac{1}{\sqrt{5}} &= \frac{1}{\sqrt{5}} \cdot \frac{\sqrt{5}}{\sqrt{5}} \\ &= \frac{\sqrt{5}}{5}.\end{aligned}$$

Ex) $\frac{3}{2\sqrt{7}}$ can be rationalized to

$$\begin{aligned}\frac{3}{2\sqrt{7}} &= \frac{3}{2\sqrt{7}} \cdot \frac{\sqrt{7}}{\sqrt{7}} \\ &= \frac{3\sqrt{7}}{2 \cdot 7} \\ &= \frac{3\sqrt{7}}{14}.\end{aligned}$$

Ex) $\frac{49}{\sqrt{7}}$ can be simplified to

$$\begin{aligned}\frac{49}{\sqrt{7}} &= \frac{49}{\sqrt{7}} \cdot \frac{\sqrt{7}}{\sqrt{7}} \\ &= \frac{49\sqrt{7}}{7} \\ &= 7\sqrt{7}.\end{aligned}$$

Ex) If $x > 0$, we can rationalize the denominator in $\frac{2}{\sqrt{2x}}$ to get

$$\begin{aligned}\frac{2}{\sqrt{2x}} &= \frac{2}{\sqrt{2x}} \cdot \frac{\sqrt{2x}}{\sqrt{2x}} \\ &= \frac{2\sqrt{2x}}{\sqrt{2x \cdot 2x}} \\ &= \frac{2\sqrt{2x}}{2x} \\ &= \frac{\sqrt{2x}}{x}.\end{aligned}$$

If there is a radical in the numerator in the denominator, we may not need to immediately rationalize. Instead, it will be useful to recall

$$\frac{\sqrt{a}}{\sqrt{b}} = \sqrt{\frac{a}{b}}.$$

Ex) $\frac{\sqrt{75}}{\sqrt{3}}$ can be simplified to

$$\begin{aligned}\frac{\sqrt{75}}{\sqrt{3}} &= \sqrt{\frac{75}{3}} \\ &= \sqrt{25} \\ &= 5.\end{aligned}$$

Example: Simplify, assuming the variables are positive:

$$\sqrt[3]{\frac{27x^6}{64y^9}}.$$

Solution: We have

$$\sqrt[3]{\frac{27x^6}{64y^9}} = \frac{\sqrt[3]{27x^6}}{\sqrt[3]{64y^9}} = \frac{3x^2}{4y^3}.$$

Example: Simplify by rationalizing the denominator:

$$\frac{7}{\sqrt[3]{4x}}.$$

Solution: Here we'll multiply the numerator and denominator by something that will clear the radical in the denominator. We have a $4x$ in the radical, so we need to multiply factors to get a perfect cube. We can multiply the 4 by 2 to get 8, and we will multiply the x by x^2. So we need to multiply by $\sqrt[3]{2x^2}$ in the numerator and the denominator:

$$\begin{aligned}\frac{7}{\sqrt[3]{4x}} &= \frac{7}{\sqrt[3]{4x}} \cdot \frac{\sqrt[3]{2x^2}}{\sqrt[3]{2x^2}} \\ &= \frac{7\sqrt[3]{2x^2}}{\sqrt[3]{8x^3}} \\ &= \frac{7\sqrt[3]{2x^2}}{2x}.\end{aligned}$$

Conjugates

The expressions

$$a + b\sqrt{c} \quad and \quad a - b\sqrt{c}$$

are called conjugates. For example, $1+\sqrt{2}$ and $1-\sqrt{2}$ are conjugates. They are special because when you multiply them together the radical disappears and they form a difference of squares:

$$(a + b\sqrt{c})(a - b\sqrt{c}) = a^2 - b^2c.$$

Ex) $(1+\sqrt{2})(1-\sqrt{2}) = 1 - 2 = -1.$

We use conjugates to rationalize the denominator if it has a sum with radicals in it. The key is to remember that we cannot change the value of a fraction. We have to multiply it by a form of 1, so we *multiply and divide by the conjugate.*

Example: Rationalize the denominator of

$$\frac{8}{7+\sqrt{3}}.$$

Solution: The conjugate of $7+\sqrt{3}$ is $7-\sqrt{3}$. If we multiply these conjugates together, we will clear the radical in the denominator:

$$\begin{aligned}\frac{8}{7+\sqrt{3}} &= \frac{8}{(7+\sqrt{3})}\cdot\frac{(7-\sqrt{3})}{(7-\sqrt{3})}\\ &= \frac{56-8\sqrt{3}}{7^2-\sqrt{3}^2}\\ &= \frac{56-8\sqrt{3}}{49-3}\\ &= \frac{56-8\sqrt{3}}{46}\\ &= \frac{2(28-4\sqrt{3})}{2(23)}\\ &= \frac{28-4\sqrt{3}}{23}.\end{aligned}$$

Example: Rationalize

$$\frac{3}{8-\sqrt{2}}.$$

Solution: The conjugate of $8-\sqrt{2}$ is $8+\sqrt{2}$. This is what will form a difference of squares in the denominator. We have

$$\begin{aligned}\frac{3}{8-\sqrt{2}} &= \frac{3}{(8-\sqrt{2})}\cdot\frac{(8+\sqrt{2})}{(8+\sqrt{2})}\\ &= \frac{24+3\sqrt{2}}{64-\sqrt{2}^2}\\ &= \frac{24+3\sqrt{2}}{64-2}\\ &= \frac{24+3\sqrt{2}}{62}.\end{aligned}$$

Example: Rationalize the denominator of

$$\frac{6}{\sqrt{7}-\sqrt{3}}.$$

Solution: The conjugate of $\sqrt{7}-\sqrt{3}$ is $\sqrt{7}+\sqrt{3}$. This is what will form a difference of squares in the denominator. We have

$$\begin{aligned}\frac{8}{\sqrt{7}-\sqrt{3}} &= \frac{8}{\sqrt{7}-\sqrt{3}}\cdot\frac{(\sqrt{7}+\sqrt{3})}{(\sqrt{7}-\sqrt{3})}\\ &= \frac{8(\sqrt{7}+\sqrt{3})}{\sqrt{7}^2-\sqrt{3}^2}\\ &= \frac{8(\sqrt{7}+\sqrt{3})}{7-3}\\ &= \frac{8(\sqrt{7}+\sqrt{3})}{4}\\ &= 2(\sqrt{7}+\sqrt{3}).\end{aligned}$$

We can leave the 2 factored, or we can distribute it if needed.

Section 7.4 Exercises

Simplify, assuming all variables are positive.

1. $\sqrt{3}\sqrt{12}$
2. $\sqrt{14}\sqrt{7}$
3. $(\sqrt[3]{9})^2$
4. $\sqrt[3]{36}\sqrt[3]{6}$
5. $(3+\sqrt{12})(2-3\sqrt{12})$

6. $(4+\sqrt{7})(\sqrt{7}+2)$

7. $(-1-\sqrt{2})(-1+\sqrt{2})$

8. $(\sqrt{3}+2\sqrt{7})(4\sqrt{3}-5\sqrt{7})$

9. $(3\sqrt{2}+2\sqrt{2x})(4\sqrt{2}-6\sqrt{2x})$

10. $(\sqrt{3}+5\sqrt{3x})(5\sqrt{3}-2\sqrt{3x})$

11. $\sqrt{3x}(\sqrt{6x}-\sqrt{12x})$

12. $\sqrt{5x}(\sqrt{10x}-x\sqrt{5x})$

13. $\sqrt[3]{4x}(\sqrt[3]{12x^4}-\sqrt[3]{6x^2})$

14. $\sqrt[3]{9x}(\sqrt[3]{3x^5}+\sqrt[3]{6x^2})$

15. $\sqrt[3]{3x}(\sqrt[3]{9x^4}-\sqrt[3]{x^2})$

16. $(\sqrt{6}+\sqrt{7})^2$

17. $(\sqrt{10}+\sqrt{5})^2$

18. $(\sqrt{8}-\sqrt{7})^2$

19. $(3\sqrt{2}+2\sqrt{3})^2$

20. $(4\sqrt{2x}-\sqrt{6x})^2$

21. $(1+\sqrt{3x})^2$

22. $(\sqrt[3]{4}-\sqrt[3]{3})^2$

23. $(\sqrt[3]{9}+\sqrt[3]{3})^2$

24. $(\sqrt[3]{2x}+\sqrt[3]{4x})^2$

25. $(\sqrt[3]{4x^2} + \sqrt[3]{8x^5})^2$

26. $(\sqrt[3]{9x^4} - \sqrt[3]{6x^2})^2$

Simplify the following expressions by rationalizing the denominator if necessary.

27. $\dfrac{3}{\sqrt{3}}$

28. $\dfrac{2}{\sqrt{6}}$

29. $\dfrac{4}{5\sqrt{2}}$

30. $\dfrac{3}{7\sqrt{12}}$

31. $\dfrac{5}{2\sqrt{10}}$

32. $\dfrac{36}{\sqrt{6}}$

33. $\dfrac{x}{2\sqrt{x}}$

34. $\dfrac{4}{2\sqrt{x^3}}$

35. $\dfrac{5}{\sqrt{x^5}}$

36. $\dfrac{\sqrt{6}}{\sqrt{3}}$

37. $\dfrac{\sqrt{3}}{\sqrt{6}}$

38. $\dfrac{\sqrt{27}}{\sqrt{6}}$

39. $\dfrac{\sqrt{26}}{\sqrt{13}}$

40. $\dfrac{\sqrt[3]{16}}{\sqrt[3]{2}}$

41. $\dfrac{\sqrt[3]{4}}{\sqrt[3]{32}}$

42. $\dfrac{\sqrt{15x^2}}{\sqrt{5x}}$

43. $\dfrac{\sqrt{14x^2y^3}}{\sqrt{7xy}}$

44. $\dfrac{2}{\sqrt{5x}}$

45. $\dfrac{3}{\sqrt{6x}}$

46. $\dfrac{\sqrt{8x^2}}{\sqrt{2x^3}}$

47. $\dfrac{1}{\sqrt[3]{6x}}$

48. $\dfrac{2}{\sqrt[3]{8x^2}}$

49. $\dfrac{1}{1+\sqrt{2}}$

50. $\dfrac{2}{\sqrt{3}+2}$

51. $\dfrac{4}{3+\sqrt{7}}$

52. $\dfrac{5}{8-\sqrt{5}}$

53. $\dfrac{2}{\sqrt{3}+\sqrt{5}}$

54. $\dfrac{3}{\sqrt{5}-\sqrt{7}}$

55. $\dfrac{5}{\sqrt{10}-\sqrt{5}}$

56. $\dfrac{2}{\sqrt{7}+\sqrt{6}}$

57. $\dfrac{\sqrt{5}+\sqrt{10}}{\sqrt{5}-\sqrt{10}}$

58. $\dfrac{x+\sqrt{y}}{x-\sqrt{y}}$

59. Looking ahead: Sometimes rationalizing the *numerator* can simplify an expression. In the following expression, rationalize the numerator. Simplify the numerator before multiplying the denominator, and cancel the resulting factors. You should end up with only expressions with x in the denominator.

$$\frac{\sqrt{x}+2}{x-4}$$

60. Repeat the same process as in the previous exercise for

$$\frac{\sqrt{4+h}-2}{h}.$$

7.5: Radical Equations

In this section we take our information about radicals and apply it to equations. In going from expressions to equations, we just remember that whatever we do to one side, we have to do to the other side. But most importantly, we have to check our answers.

The way to solve a radical equation is to *isolate the radical*, and then lift it by squaring both sides (or whatever root we may be dealing with). Again, we have to be extremely careful to *check the answer*, because we could be introducing what are called extraneous solutions. These are extra solutions that may solve a version of the equation we've modified but do *not* satisfy the original equation.

Example: Solve $\sqrt{x-1}=3$.

Solution: The radical has been isolated, so we can square both sides:

$$(\sqrt{x-1})^2 = 3^2$$
$$x-1=9$$
$$x=10$$

A check of our work shows that $\sqrt{10-1}=\sqrt{9}=3$, so the equation is satisfied.

Example: Solve $1+\sqrt{2x}=8$.

Solution: We isolate the radical, so we have

$$\begin{aligned}1+\sqrt{2x}&=8\\ \sqrt{2x}&=7\\ (\sqrt{2x})^2&=7^2\\ 2x&=49\\ x&=\frac{49}{2}.\end{aligned}$$

Plugging this back in to check, we get

$$\begin{aligned}1+\sqrt{2\cdot\frac{49}{2}}&=?8\\ 1+\sqrt{49}&=?8\\ 1+7&=8\end{aligned}$$

Our solution is $x=\frac{49}{2}$.

Example: Solve $\sqrt{3x+1}+8=1$.

Solution: Isolate the radical by subtracting 8 from both sides. Then square both sides:

$$\begin{aligned}\sqrt{3x+1}+8&=1\\ \sqrt{3x+1}&=-7\\ \left(\sqrt{3x+1}\right)^2&=(-7)^2\\ 3x+1&=49\\ 3x&=48\\ x&=\frac{48}{3}=16.\end{aligned}$$

To check our work, let's plug in $x = 16$ into the original equation:

$$\sqrt{3\cdot(16)+1}+8=?1$$
$$\sqrt{49}+8=?1$$
$$7+8\neq 1$$

The answer $x = 16$ does not satisfy the equation. This means there are no solutions to the equation! In the second step of solving the equation, we had

$$\sqrt{3x+1}=-7.$$

Thinking about the expression $\sqrt{3x+1}$, we realize that square roots must always be nonnegative. So there is no way that $\sqrt{3x+1}$ could be a negative number.

There are two other types of equation that we will deal with here: one where we'll use roots *besides* square roots, and one where we'll have multiple roots.

Example: Solve $\sqrt[3]{3x+5}=-4$.

Solution: Since we have a cube root of x, it would make sense to cube both sides:

$$\left(\sqrt[3]{3x+5}\right)^3=(-4)^3$$
$$3x+5=-64$$
$$3x=-69$$
$$x=-23.$$

Plugging this back in, we have

$$\sqrt[3]{3(-23)+5}=?-4$$
$$\sqrt[3]{-69+5}=?-4$$
$$\sqrt[3]{-64}=-4.$$

Thus, $x = -23$ is the solution.

If there is more than one radical, then we isolate one radical and square both sides. This will not get rid of all the radicals, however, so we have to isolate again.

Example: Solve

$$\sqrt{x} - \sqrt{x-3} = 1.$$

Solution: Moving one of the radicals to the other side will make the solution easier to find. If we add $\sqrt{x-3}$ to both sides, we have

$$\begin{aligned}
\sqrt{x} &= 1 + \sqrt{x-3} \\
\left(\sqrt{x}\right)^2 &= \left(1 + \sqrt{x-3}\right)^2 \\
x &= \left(1 + \sqrt{x-3}\right)\left(1 + \sqrt{x-3}\right).
\end{aligned}$$

Make sure to FOIL the right-hand side! We have

$$\begin{aligned}
x &= 1 + \sqrt{x-3} + \sqrt{x-3} + \left(\sqrt{x-3}\right)^2 \\
x &= 1 + 2\sqrt{x-3} + x - 3 \\
x &= -2 + 2\sqrt{x-3} + x \\
0 &= -2 + 2\sqrt{x-3} \\
2 &= 2\sqrt{x-3} \\
1 &= \sqrt{x-3} \\
1^2 &= \left(\sqrt{x-3}\right)^2 \\
1 &= x - 3 \\
4 &= x.
\end{aligned}$$

Checking $x = 4$, we have

$$\sqrt{4} - \sqrt{4-3} = ?1$$
$$2 - \sqrt{1} = ?1$$
$$2 - 1 = 1.$$

So $x = 4$ is the solution.

Section 7.5 Exercises

Solve each equation. Remember to check any possible solutions.

1. $\sqrt{5x} = 10$
2. $\sqrt{5x} = -1$
3. $-\sqrt{6x} = -3$
4. $-\sqrt{3x} + 4 = 10$
5. $3 + \sqrt{4x} = 6$
6. $\sqrt{2x+3} = 4$
7. $\sqrt{2x-3} + 6 = 10$
8. $\sqrt{10x-4} + 7 = 20$
9. $\sqrt{14+x^2} = x$
10. $3 = 5 - 2\sqrt{1-x}$
11. $3\sqrt{4+x} - 6 = 9$
12. $2 = 4\sqrt{x-7} - 10$

13. $4+\sqrt[3]{x}=7$

14. $2-\sqrt[3]{x}=-1$

15. $\sqrt[3]{3x+2}-5=-4$

16. $\sqrt[3]{1-4x}-2=-5$

17. $2\sqrt[3]{2x+1}-5=3$

18. $\sqrt{2x+5}=1+\sqrt{2x+4}$

19. $\sqrt{x+3}-1=\sqrt{x+1}$

20. $\sqrt{x+2}+\sqrt{x+4}=1$

21. $\sqrt{3x+1}-1=\sqrt{3x-4}$

22. $\sqrt{4x-2}+3-\sqrt{4x-5}=0$

7.6: Complex Numbers

Taking the square root of negative numbers leads to imaginary numbers, which are nonreal numbers.

We define

$$i=\sqrt{-1},$$

which is sometimes called the imaginary unit. In Section 7.7, we will begin to solve more complicated equations that may have imaginary numbers as their solutions. But first we need to know what it means for an equation to have real solutions as opposed to imaginary solutions. (In later courses we find out what imaginary solutions will look like in a graph of an equation.)

Recall Section 1.1, where we examined real numbers as the largest set of numbers we were dealing with. Now we'll make that set larger by introducing the idea of complex numbers. Complex numbers have the form

$$a + bi,$$

where $a, b \in \mathbb{R}$ and $i = \sqrt{-1}$. The real number a is called the real part of the complex number $a + bi$. The real number b is called the imaginary part of the complex number $a + bi$.

Ex) $2 + 3i$ is a complex number. The real part of $2 + 3i$ is 2, and the imaginary part is 3.

Ex) $-18 - \sqrt{2}i$ is a complex number. The real part is -18, and the imaginary part is $-\sqrt{2}$.

Ex) 3 is a complex number, because it can be written as $3 + 0i$.

See Figure 7-1 below. Remember any higher set of numbers is automatically one that is connected below it.

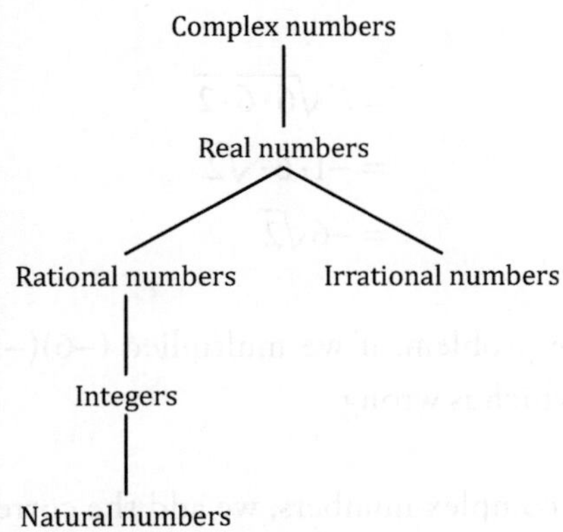

Fig. 7-1

So 3 is a complex number, a real number, a rational number, an integer, and a natural number.

In this section we will deal with some algebra of the complex numbers. The main rule to remember is that *if we encounter the square root of a negative number, we remove the i first.*

This is very important:

$$\sqrt{a}\sqrt{b} \neq \sqrt{ab} \text{ if } a \text{ and } b \text{ are negative.}$$

Ex) $\sqrt{-4} = \sqrt{-1 \cdot 4} = i \cdot 2 = 2i$

Ex) $\sqrt{-2} = \sqrt{-1 \cdot 2} = i\sqrt{2}$

Multiplication of imaginary numbers is done by removing the i first.

Example: Simplify $\sqrt{-6}\sqrt{-12}$.

Solution: We must remove the i first, before we multiply the numbers under the radical:

$$\sqrt{-6}\sqrt{-12} = i\sqrt{6} \cdot i\sqrt{12}$$

Now 6 and 12 are positive, so we can multiply them under the radical:

$$\begin{aligned} &= i^2\sqrt{6 \cdot 6 \cdot 2} \\ &= -1 \cdot 6 \cdot \sqrt{2} \\ &= -6\sqrt{2} \end{aligned}$$

Note: At the beginning of the problem, if we multiplied (−6)(−12) under a single radical we would get a positive number, which is wrong.

When adding and subtracting complex numbers, we add the corresponding real parts and their corresponding imaginary parts. Again, think of this as being similar to adding like terms.

Example: Simplify $(3-4i)+(15-100i)$.

Solution: We have

$$\begin{aligned} (3-4i)+(15-100i) &= 3+15-4i-100i \\ &= 18-104i. \end{aligned}$$

Example: Simplify $(\sqrt{2}-4i)-(-\sqrt{2}-4i)$.

Solution: Being careful to distribute the negative, we have

$$\begin{aligned}(\sqrt{2}-4i)-(-\sqrt{2}-4i) &= \sqrt{2}-4i-(-\sqrt{2})-(-4i)\\ &= \sqrt{2}-4i+\sqrt{2}+4i\\ &= 2\sqrt{2}.\end{aligned}$$

The *i* Cycle

We will use the $i=\sqrt{-1}$ notation often. When multiplying and dividing, we will encounter powers of i, which is a sequence that cycles back every four powers.

Ex) Think of squaring i as lifting the radical. Since $i=\sqrt{-1}$, we have

$$i^2=-1$$

Ex) i^3 can be written as $i^2\cdot i$, and since $i^2=-1$, we have

$$i^3=i^2\cdot i=-1\cdot i=-i.$$

Ex) Finally, i^4 can be written as $i^2\cdot i^2=-1\cdot -1$, which is 1.

Example: Simplify i^5+i^{12}.

Solution: Since $i^4=1$, we figure out what i^5 is by multiplying both sides by i:

$$\begin{aligned}i^4&=1\\ i^4\cdot i&=1\cdot i\\ i^5&=i.\end{aligned}$$

Any further powers will repeat. In fact, we have

$$\begin{aligned}&i^1=i, i^2=-1, i^3=-i, i^4=1\\ &i^5=i, i^6=-1, i^7=-i, i^8=1\end{aligned}$$

The powers cycle back to i after every fourth power. If the power is divisible by 4, the result will be 1. So $i^5 + i^{12} = i + 1$.

Example: What is i^{97}?

Solution: Since the powers of i cycle back every four powers, we know that if the power is divisible by 4, the result is 1. We can write i^{97} as the product of i^{96} and i. And 96 is divisible by 4, so $i^{96} = 1$. So we have

$$\begin{aligned} i^{97} &= i^{96} \cdot i \\ &= 1 \cdot i \\ &= i. \end{aligned}$$

Multiplication of Complex Numbers

We multiply complex numbers using the distributive property, or FOILing.

Example: Simplify $(8-9i)(4+3i)$.

Solution: We FOIL this for

$$\begin{aligned} (8-9i)(4+3i) &= 32 + 24i - 36i - 27i^2 \\ &= 32 - 12i - 27(-1) \\ &= 32 + 27 - 12i \\ &= 59 - 12i. \end{aligned}$$

Example: Simplify $(2i+\sqrt{7})(5i-\sqrt{7})$.

Solution: We FOIL this for

$$\begin{aligned} (2i+\sqrt{7})(5i-\sqrt{7}) &= 2i \cdot 5i - 2i\sqrt{7} + \sqrt{7} \cdot 5i - \sqrt{7} \cdot \sqrt{7} \\ &= 10i^2 + 3i\sqrt{7} - 7 \\ &= -10 - 7 + 3i\sqrt{7} \\ &= -17 + 3i\sqrt{7}. \end{aligned}$$

Dividing Complex Numbers

In Section 7.4, we dealt with conjugates. The complex conjugate of $a + bi$ is $a - bi$.

Having an i in the denominator is like having a radical in the denominator. It may help to simplify our expressions by rationalizing the denominator.

Example: Simplify $\frac{2}{i}$ by rationalizing the denominator.

Solution: To get rid of the radical in the denominator, we multiply by $\frac{i}{i}$. We have

$$\begin{aligned}\frac{2}{i} &= \frac{2}{i} \cdot \frac{i}{i} \\ &= \frac{2i}{i^2} \\ &= \frac{2i}{-1} \\ &= -2i.\end{aligned}$$

Example: Simplify $\frac{5}{3+4i}$ by rationalizing the denominator.

Solution: The conjugate of $3 + 4i$ is $3 - 4i$. So we multiply and divide by $3 - 4i$:

$$\begin{aligned}\frac{5}{3+4i} &= \frac{5}{(3+4i)} \cdot \frac{(3-4i)}{(3-4i)} \\ &= \frac{5(3-4i)}{9-16i^2} \\ &= \frac{5(3-4i)}{9-16(-1)} \\ &= \frac{5(3-4i)}{9+16} \\ &= \frac{5(3-4i)}{25} \\ &= \frac{3-4i}{5}.\end{aligned}$$

If we want to write this in the form $a + bi$, we can split the fraction and get $\frac{3}{5} - \frac{4}{5}i$.

Example: Simplify by rationalizing the denominator:

$$\frac{4-2i}{2+3i}.$$

Solution: We multiply and divide by the complex conjugate of the denominator, which is $2-3i$. We have

$$\begin{aligned}\frac{4-2i}{2+3i} &= \frac{(4-2i)}{(2+3i)}\cdot\frac{(2-3i)}{(2-3i)}\\ &= \frac{8-12i-4i+6i^2}{4-9i^2}\\ &= \frac{8-16i-6}{4+9}\\ &= \frac{2-16i}{13}.\end{aligned}$$

We can leave the answer like this, or we canwrite it in the form of $a+bi$ by splitting the fraction for $\frac{2}{13}-\frac{16}{13}i$.

Section 7.6 Exercises

Write the following numbers in the form $a+bi$, where $a,b\in\mathbb{R}$.

1. $\sqrt{-9}$
2. $\sqrt{-40}$
3. $\sqrt{-18}+\sqrt{-32}$
4. $\sqrt{-27}+\sqrt{-300}$
5. $\sqrt{-\frac{49}{25}}$

6. $\sqrt{\frac{-49}{-25}}$

7. $\frac{\sqrt{-49}}{\sqrt{-25}}$

8. $2-\sqrt{-2}$

9. $3+\sqrt{-8}$

10. $\sqrt{-8}\sqrt{-6}$

11. $\sqrt{-12}\sqrt{-12}$

12. $\sqrt{-13}\sqrt{-26}$

13. $\sqrt{-4}\sqrt{-9}$

14. $\sqrt{-7}\sqrt{-28}$

15. $(2+3i)-(4+2i)$

16. $(2+3i)-(4-2i)$

17. $(0.1+0.2i)+(0.6-i)$

18. $\left(\frac{1}{2}+\frac{\sqrt{3}}{2}i\right)-\left(\frac{1}{2}-\frac{\sqrt{3}}{2}i\right)$

19. $\left(\frac{\sqrt{2}}{2}+\frac{\sqrt{2}}{2}i\right)+\left(\frac{\sqrt{2}}{2}-\frac{\sqrt{2}}{2}i\right)$

20. $i+i^2+i^3$
21. $i-i^2$
22. $i^3+i^7-i^9$
23. i^{74}
24. i^{132}

25. i^{102}
26. $i^{140} + i^{70}$
27. $2i^{79} + i^{82}$
28. $4i(2i)$
29. $2i(3+2i)$
30. $i(1-i)$
31. $-i(-1-i)$
32. $(-1+3i)(1+2i)$
33. $(4+3i)(4-3i)$
34. $(\sqrt{3}+2i)(2\sqrt{3}-3i)$
35. $(\sqrt{3}+\sqrt{5}i)(2\sqrt{3}-3\sqrt{5}i)$
36. $(\sqrt{6}+\sqrt{7}i)(\sqrt{6}-\sqrt{7}i)$
37. $\frac{7}{i}$
38. $\frac{3}{i^3}$
39. $-\frac{2}{i}$
40. $\frac{2}{3-i}$
41. $\frac{1}{1+i}$
42. $\frac{\sqrt{2}}{\sqrt{2}+i}$
43. $\frac{3+2i}{3-2i}$
44. $\frac{1-i}{6+8i}$
45. $\frac{2+i}{5-12i}$

7.7: VARIATION

In this section we'll handle some applications that involve radicals. One important concept is variation.

We say that y varies directly (or is directly proportional) as x if there exists a nonzero constant k such that

$$y = kx.$$

The constant k is called the constant of proportionality.

Example: Suppose p varies directly as q. If $p = 40$ when $q = 10$, then what is p when $q = 30$?

Solution: Since p is directly proportional to q, we have

$$p = kq$$

for some fixed k. To determine what k is, we use the other information given in the problem. When $q = 10$, we have $p = 40$. Plugging this into $p = kq$, we have

$$\begin{aligned} 40 &= k(10) \\ 4 &= k. \end{aligned}$$

So now we know that $p = 4q$. Finally, we are asked what p is when $q = 30$. We have

$$\begin{aligned} p &= 4(30) \\ &= 120. \end{aligned}$$

We say that y is inversely proportional to x if there is a nonzero constant k such that

$$y = \frac{k}{x}.$$

Example: If v is inversely proportional to $\sqrt{w}$, and v is 2 when w is 16, then what is v when $w = 18$?

Solution: We have

$$v = \frac{k}{\sqrt{w}}.$$

When $v = 2$ and $w = 16$, we are able to find the constant of proportionality:

$$\begin{aligned} 2 &= \frac{k}{\sqrt{16}} \\ 2 &= \frac{k}{8} \\ 16 &= k. \end{aligned}$$

So now we know that

$$v = \frac{16}{\sqrt{w}}.$$

What we are looking for is the value of v when $w = 18$:

$$\begin{aligned} v &= \frac{16}{\sqrt{18}} \\ &= \frac{16}{\sqrt{9 \cdot 2}} \\ &= \frac{16}{3\sqrt{2}} \\ &= \frac{16}{3\sqrt{2}} \cdot \frac{\sqrt{2}}{\sqrt{2}} \\ &= \frac{16\sqrt{2}}{3 \cdot 2} \\ &= \frac{8\sqrt{2}}{3}. \end{aligned}$$

Joint variation is used when a variable varies directly with some variables and inversely as other variables. For example, if we want to say that y varies directly as x and inversely as z, we would say

$$y = \frac{kx}{z}.$$

Example: The kinetic energy of an object varies jointly with the product of its mass and the square of its velocity. If a 10-kg object moving at $4\frac{\text{m}}{\text{s}}$ has a kinetic energy of 80 J $\left(1\text{J}=1\text{kg}\frac{\text{m}^2}{\text{s}^2}\right)$, then how much kinetic energy does a 15-kg object moving at $3\frac{\text{m}}{\text{s}}$ have?

Solution: If E represents the kinetic energy, m the mass, and v the velocity, then

$$E = kmv^2.$$

We have the information that $m = 10$, $v = 4$, $E = 80$. We use it to find the constant k:

$$\begin{aligned} 80 &= k(10)(4)^2 \\ 80 &= k \cdot 10 \cdot 16 \\ 80 &= 160k \\ \frac{1}{2} &= k, \end{aligned}$$

and our equation becomes

$$E = \frac{1}{2}mv^2.$$

We use this to find the energy when $m = 15$ and $v = 3$:

$$\begin{aligned} E &= \frac{1}{2}(15)(3)^2 \\ &= \frac{1}{2} \cdot 15 \cdot 9 \\ &= \frac{135}{2}, \end{aligned}$$

or 67.5 J.

Section 7.7 Exercises

1. If y varies directly as b, and y is $\frac{1}{2}$ when b is $\frac{1}{4}$, then what is the value of y when $b = 1$?
2. Suppose q varies inversely as x. When $x = 15$, $q = \frac{1}{30}$. What is q when $x = 20$?
3. If l varies jointly with s and t, and $l = 60$ when $s = 2$ and $t = 3$, then what is the value of l when $s = 10$ and $t = 7$?
4. Suppose v varies directly as w and inversely as z. When $w = 3$ and $z = 2$, v is 30. What is the value of z when $v = 10$ and $w = 7$?
5. Suppose q varies directly as a and inversely with the square root of b. If $a = 3$ and $b = 6$, then $q = 10$. What is the value of q when $a = \sqrt{2}$ and $b = 8$?
6. Suppose j varies inversely with the product of b and c^2. When $b = 4$ and $c = 6$, $j = 3$. What is the value of j when b is 10 and c is $\sqrt{12}$?
7. If x varies directly with the cube of y and inversely with the square of t, and x is 15 when y is 2 and t is 3, what is the value of x when y is 3 and t is 4?
8. If w varies directly with the product of q and z, and w is 0.3 when $q = 0.25$ and $z = 0.1$, then what is the value of z when w is 1.4 and q is 2.7?
9. Hooke's law states that the force required to stretch a spring x meters beyond its natural length is directly proportional to x. If in a certain spring a force of 20 N (newtons) is required to stretch a spring 5 m beyond its natural length, what force is required to stretch the spring 6 m beyond its natural length?
10. Newton's law of universal gravitation states that the force between two objects is directly proportional to the product of their masses m_1 and m_2 while also inversely proportional to the square of the distance between them. If there is an 8-N force between two objects with masses 10 kg and 20 kg that are 3 m apart, then what force is between two objects with masses 8 and 4 kg that are 5 m apart?
11. The frequency of a sine function is inversely proportional to its period. If the frequency of a sine function with period 2π is $\frac{1}{2\pi}$, what is the frequency of a sine function with period $\frac{\pi}{2}$?
12. Boyle's law states that the volume of a gas held at a constant temperature is inversely proportional to the applied pressure (assuming the mass is constant). The unit name for pressure is **torr**. If a certain gas has an initial volume of 250 mL at a pressure of 380 Torr, what is the pressure when the volume is compressed to 200 mL?
13. The number of people at a party is directly proportional to the number of hours that have elapsed since the party started. If a party was scheduled to begin at 7:00 p.m., suppose that there were 9 people there at 8:00 p.m. How many people were at the party at midnight?

14. The slope of a line is directly proportional to its rise and inversely proportional to its run. If a certain line with rise 8 and run $\frac{1}{2}$ has slope 16, then what is the slope of a line with rise $\frac{1}{3}$ and run $\frac{2}{3}$?

Chapter 7 Review

7-A

1. Simplify $4^{\frac{1}{2}} + 81^{\frac{3}{2}}$.
2. Divide

$$\frac{4x^{\frac{5}{6}}}{2x^{\frac{1}{6}}}.$$

3. Factor $x^{\frac{3}{2}} + x^{\frac{1}{2}}$.

4. Simplify $\sqrt{28}$.

5. Simplify $\sqrt{z^9 p^8}$, assuming variables are positive.

6. Simplify $\sqrt{28} + \sqrt{56}$.

7. Simplify $\sqrt{8x} + \sqrt{162x}$, assuming $x > 0$.

8. Multiply and simplify $(3+\sqrt{2})(3-\sqrt{2})$.
9. Rationalize the denominator:

$$\frac{1}{\sqrt{6}}.$$

10. Rationalize the denominator:

$$\frac{1}{2+\sqrt{3}}.$$

11. Solve $\sqrt{4x-5}+1=10$.
12. Simplify $\sqrt{-4}\sqrt{-5}$.
13. Multiply out and simplify $(2-5i)^2$.
14. Rationalize the denominator:

$$\frac{2}{1+i}.$$

15. If y varies directly as p^2, and y is 36 when p is 3, what is the value of y when $p=13$?

7-B

1. Simplify $100^{\frac{3}{2}}+100^{-\frac{1}{2}}$.
2. Simplify

$$\frac{8x^{\frac{3}{4}}y^{\frac{5}{3}}}{6x^{\frac{5}{4}}y^{\frac{2}{3}}}.$$

3. Factor to simplify

$$(2x-1)^{\frac{1}{2}}+x(2x-1)^{-\frac{1}{2}}.$$

4. Simplify $\sqrt{700}$.

5. Simplify $\sqrt[3]{54x^{10}y^{22}}$.

6. Simplify $\sqrt{72}+\sqrt{128}-\sqrt{242}$.

7. Simplify $\sqrt[3]{128x}-\sqrt[3]{24x}$.

8. Multiply out and simplify $(4-5\sqrt{7})(3+2\sqrt{7})$.
9. Rationalize the denominator:

$$\frac{3}{\sqrt{32}}.$$

10. Rationalize the denominator:

$$\frac{1+\sqrt{7}}{3-2\sqrt{7}}.$$

11. Solve the equation $3\sqrt{8-2x}-5=7$.

12. Simplify $\sqrt{-10}\sqrt{-15}$.

13. Multiply out and simplify $(\sqrt{6}+5i)(3\sqrt{6}-3i)$.
14. Rationalize the denominator

$$\frac{8-i}{6-8i}.$$

15. Assume that b varies directly as a and inversely as t^3. When $a=3$ and $t=2$, $b=10$. What is the value of b when a is 4 and t is 3?

CHAPTER 8

Quadratics and Functions

This chapter deals with second-degree polynomials. We will talk about their equations, their graphs, and their inequalities. Much of what talked about in previous chapters will tie in, including factoring, radicals, and solving inequalities.

8.1: Solving Quadratic Equations and the Square-Root Property

This section addresses the solutions to

$$ax^2 + bx + c = 0,$$

where $a, b, c \in \mathbb{R}$ and $a \neq 0$. (Note that if $a = 0$, this is a linear equation.) Quadratic equations have many applications, including projectile motion and reflection of sound waves.

First, we will handle the case when $b = 0$. If we solve an equation that looks like

$$ax^2 + c = 0,$$

we isolate the x as usual. But now we use the square-root property:

$$\text{If } x^2 = p, \text{ then } x = \pm\sqrt{p}.$$

A word on notation: The notation $\pm\sqrt{p}$ is shorthand for $\sqrt{p}$ and $-\sqrt{p}$.

So if $x = -\sqrt{p}$, then $x = +\sqrt{p}$ (which is just the square root of p) or $x = -\sqrt{p}$.

Let's think about why the square-root property makes sense.

Ex) If we solve $x^2 = 4$, we are asking "What numbers, when squared, are 4?" Certainly, $2^2 = 4$, so $x = 2$ is a solution. But it is also true that $(-2)^2 = (-2)(-2) = 4$. There are two solutions to this equation: the positive and negative square roots of 4. Thus, if

$$x^2 = 4,$$

then

$$\begin{aligned} x &= \pm\sqrt{4} \\ &= \pm 2. \end{aligned}$$

Sometimes the positive square root of a number is called the principal square root of a number. This is because in conversational English, if someone asks what the square root of 9 is, we say 3, even though it is also true that $(-3)^2 = 9$.

Example: Solve $3x^2 = 9$.

Solution: We have

$$\begin{aligned} 3x^2 &= 9 \\ x^2 &= 3 \\ x &= \pm\sqrt{3}. \end{aligned}$$

Example: Solve $4t^2 - 22 = 8$.

Solution: We have

$$\begin{aligned} 4t^2 - 22 &= 8 \\ 4t^2 &= 30 \\ t^2 &= \frac{30}{4} = \frac{15}{2} \\ t &= \pm\sqrt{\frac{15}{2}}. \end{aligned}$$

We can rationalize this if we want:

$$\begin{aligned} t &= \pm\frac{\sqrt{15}}{\sqrt{2}} \cdot \frac{\sqrt{2}}{\sqrt{2}} \\ t &= \pm\frac{\sqrt{30}}{2}. \end{aligned}$$

Example: Solve $14x^2 - 7 = -7$.

Solution: We have

$$\begin{aligned} 14x^2 &= 0 \\ x^2 &= 0 \\ x &= \pm\sqrt{0} \\ x &= 0. \end{aligned}$$

This is an example of a quadratic equation with one *distinct* solution.

Example: Solve $x^2 + 5 = 11$.

Solution: We have

$$\begin{aligned} x^2 + 11 &= 5 \\ x^2 &= -6 \\ x &= \pm\sqrt{-6} \\ x &= \pm i\sqrt{6}. \end{aligned}$$

Equations of the Form $ax^2 + bx + c = 0$ or $ax^2 + bx = 0$

The remainder of this section and the next two sections are devoted to solving this equation:

$$x^2 + bx + c = 0 \quad or \quad ax^2 + bx = 0$$

In all of our previous work, we were isolating x or isolating radicals of x. Here we actually want everything on the same side of the equation, because that's how we can factor it (or that's how we can use some of the formulas that will be developed in the next two sections).

In this section, we address instances where $ax^2 + bx + c$ can be factored. We already handled factoring these expressions in Chapter 5. Now, when solving the equations, we need one more step, called the zero-product property:

When two numbers multiply together to get zero, one of them must be zero.
That is, if $ab = 0$, then either $a = 0$ or $b = 0$.

This should make sense: If two numbers are multiplied together and the result is zero, then one of them has to be zero.

Note that this is not the case for any number other than zero. If two numbers multiply together to get 12, one of the numbers does not have to be 12. (For example, $\frac{1}{2}$ and 24 multiply together to get 12, but neither of the numbers is 12.)

This is how we will solve a factored quadratic expression set equal to zero. If we have

$$(x-a)(x-b)=0,$$

then either $x-a=0$ or $x-b=0$.

Example: Solve $x^2-5x=0$.

Solution: When we factor out the GCF, which is x, we have

$$x(x-5)=0,$$

so either $x=0$ or $x-5=0$. Thus, the solutions are $x=0$ and $x=5$.

Example: Solve $36x^2-81=0$.

Solution: We can use the square-root property as above, or we can factor this as a difference of squares. If we recognize the difference of squares, this becomes

$$(6x+9)(6x-9)=0.$$

So

$$\begin{aligned} 6x+9=0 \quad &or \quad 6x-9=0 \\ x=-\frac{9}{6} \quad &or \quad x=\frac{9}{6} \\ x=-\frac{3}{2}, \quad &\quad \frac{3}{2}. \end{aligned}$$

Example: Solve $2x^2 - 7x - 4 = 0$.

Solution: Let's try to factor this. We have

$$(2x+1)(x-4)=0,$$

so $2x+1=0$ or $x-4=0$. Our solutions are $x=-\frac{1}{2}$ or $x=4$.

Example: Solve $9x^2 + 12x + 4 = 0$.

Solution: If we factor the trinomial, we have choices of $3x$ and $3x$ or $9x$ and x. We use the $3x$ pair because $9x$ will be too high. Our factorization is

$$(3x+2)(3x+2)=0,$$

which solves to $3x+2=0$ or $3x+2=0$. So this equation has only one unique solution:

$$x=-\frac{2}{3}.$$

Example: Solve $x^2 + 6x = 7$.

Solution: It is important to note that we should not start by factoring the x out of the left-hand side. We did that when solving $x^2 - 5x = 0$, because the right-hand side was zero. But we cannot factor $x^2 + 6x = 7$ to get $x(x+6)=7$ (this is a legal move) and then set x equal to 7 and $x+6$ equal to 7 (this is an *illegal* move). We must factor this like a quadratic expression. We move the constant to the same side as the x's:

$$\begin{aligned} x^2+6x-7&=0 \\ (x+7)(x-1)&=0 \\ x+7=0 \quad \text{or} \quad x-1&=0 \\ x=-7 \quad \text{or} \quad x&=1 \end{aligned}$$

Example: Solve $\frac{35}{x} - x = 2$.

Solution: Recall from Section 6.3 that we multiply both sides by x (the LCD) to clear the fractions:

$$x\left(\frac{35}{x} - x\right) = x \cdot 2$$
$$x \cdot \frac{35}{x} - x^2 = 2x$$
$$35 - x^2 = 2x$$
$$35 = x^2 + 2x$$
$$0 = x^2 + 2x - 35$$
$$0 = (x-5)(x+7)$$
$$x - 5 = 0 \quad or \quad x + 7 = 0$$
$$x = 5 \quad or \quad x = -7$$

We need to test these solutions. If $x = 5$, we have

$$\frac{35}{5} - 5 = ?2$$
$$7 - 5 = 2.$$

Testing $x = -7$ yields

$$\frac{35}{-7} - (-7) = ?2$$
$$-5 + 7 = ?2$$
$$2 = 2.$$

So both $x = 5$ and $x = -7$ are solutions.

A Note about Radical Equations

In Section 7.6, we would solve equations like $\sqrt{x-1} = 3$ by squaring both sides.

Now we can really understand why it is important to check the solutions to radical equations. Suppose we are solving $x = 1$, which has only one solution. Now suppose we square both sides, which gives us

$$x^2 = x.$$

To solve this, we have to move x to the same side to get $x^2 - x = 0$. If we factor this, we see that $x(x-1) = 0$, which has solutions $x = 0$ or $x = 1$. Squaring both sides to any equation may introduce extra solutions.

Example: Solve $t - \sqrt{t-3} = 5$.

Solution: Isolate the radical first:

$$\begin{aligned} -\sqrt{t-3} &= 5 - t \\ \sqrt{t-3} &= t - 5 \\ \left(\sqrt{t-3}\right)^2 &= (t-5)^2. \end{aligned}$$

Now FOIL the right-hand side:

$$\begin{aligned} t - 3 &= t^2 - 10t + 25 \\ t^2 - 11t + 28 &= 0 \\ (t-4)(t-7) &= 0 \\ t - 4 &= 0 \quad or \quad t - 7 = 0. \end{aligned}$$

So $t = 4, 7$ are the possible solutions. Let's plug $t = 4$ into the original equation:

$$\begin{aligned} 4 - \sqrt{4-3} &= ?5 \\ 4 - 1 &\neq 5. \end{aligned}$$

So $t = 4$ is not a solution. Let's try $t = 7$:

$$7 - \sqrt{7-3} = ?5$$
$$7 - \sqrt{4} = ?5$$
$$7 - 2 = 5$$

So $t = 7$ is the only solution.

Section 8.1 Exercises

Solve the following equations.

1. $y^2 = 7$
2. $2x^2 = 3$
3. $-2t^2 = 3$
4. $\frac{1}{4}x^2 - 4 = 0$
5. $11x^2 = 1$
6. $x^2 = 0.01$
7. $\frac{1}{2}x^2 + \frac{3}{4} = 0$
8. $x^2 + 18 = 20$
9. $y^2 - 18 = 20$
10. $w^2 + 2w = 0$
11. $p^2 = 2p$
12. $2r^2 = r$
13. $x^2 - 2x - 3 = 0$
14. $2z^2 - 3z - 2 = 0$
15. $u^2 - 49 = 0$
16. $x^2 + x - 42 = 0$
17. $m^2 - m = 42$
18. $x^2 - 8x = 0$
19. $x^2 - 8x + 12 = 0$

20. $x+10+\frac{21}{x}=0$

21. $x+1=\frac{20}{x}$

22. $x-\frac{1}{x}=0$

23. $15=2x-x^2$

24. $2\sqrt{y+4}-1=\sqrt{2y+9}$

25. $\sqrt{3v+1}-\sqrt{v-4}=3$

26. $\sqrt{x}+\sqrt{2x}=1$

27. $\sqrt{2x}+\sqrt{3x}=2$

28. $\sqrt{2x+3}=\sqrt{x+1}-\sqrt{x+4}$

29. $\sqrt{4x+1}+\sqrt{x}=2$

8.2: Completing the Square and Factoring

Completing the square is a way to force any quadratic equation to use the square-root property. We like equations that don't have a linear term; for example, solving $3x^2=7$ is a lot easier than solving $3x^2+6x+7=0$. Completing the square is a transformation that makes such equations more manageable.

Below, we'll present an example to show how to complete the square. Then we'll describe the general process.

Consider the quadratic equation $x^2-6x+2=0$. To solve it, first we move the constant to the other side, to get

$$x^2-6x=-2.$$

Now, we take half of the linear coefficient and square it:

$$\left(\frac{-6}{2}\right)^2.$$

This simplifies to $(-3)^2 = 9$.

Next, we add 9 to both sides of the equation. This may seem strange, but because this is an equation it is legal to add any number we like as long as we add it to both sides. By adding 9, we get

$$x^2 - 6x + 9 = -2 + 9$$
$$x^2 - 6x + 9 = 7.$$

The left-hand side is now what's called a perfect square trinomial, which will not only be easy to factor, but also will have the same binomial for both factors. Indeed, we see

$$(x-3)(x-3) = 7.$$

We write this as

$$(x-3)^2 = 7.$$

The beauty of this is that *now we can use the square-root property*. We take plus/minus the square root of both sides to get

$$x - 3 = \pm\sqrt{7},$$

then we add 3 to both sides:

$$x = 3 \pm \sqrt{7}.$$

This gives us the solutions

$$x = 3 - \sqrt{7}, \quad or \quad x = 3 + \sqrt{7}.$$

Here are the steps for completing the square:

1. Make sure that the constant is on one side of the equation, and the terms with x are on the other side.
2. Never complete the square unless the coefficient of the quadratic term is 1. If it is not 1, factor the side that has the x's.
3. Take *half of the linear coefficient, then square it, then add it to both sides.* Be careful if the quadratic term had a coefficient other than 1.
4. One side should be a perfect square trinomial and should factor into $(ax+b)^2$.
5. Take plus/minus the square root of both sides, and then solve for x.

Example: Solve $3x^2 + 6x - 1 = 0$.

Solution: Following the steps above, we need to move the −1 to the other side. Then, since the quadratic term has a coefficient, we factor it out:

$$3x^2 + 6x = 1$$

$$3(x^2 + 2x) = 1$$

$$3(x^2 + 2x + \_\_) = 1 + \_\_.$$

At this step, we take half the linear coefficient, which is $\frac{2}{2}$, or 1. If we square this, we also get 1. Now, we put a 1 in the blank on the left:

$$3(x^2 + 2x + 1) = 1 + \_\_$$

But we do not add 1 to the other side. Because of the quadratic coefficient, we are really adding $3 \cdot 1$ to both sides. So we have

$$
\begin{aligned}
3(x^2+2x+1) &= 1+3 \\
3(x+1)^2 &= 4 \\
(x+1)^2 &= \frac{4}{3} \\
x+1 &= \pm\sqrt{\frac{4}{3}} \\
x+1 &= \pm\frac{2}{\sqrt{3}} \\
x+1 &= \pm\frac{2\sqrt{3}}{3} \\
x &= -1\pm\frac{2\sqrt{3}}{3}.
\end{aligned}
$$

Our solutions are $x = -1 - \frac{2\sqrt{3}}{3}$ or $x = -1 + \frac{2\sqrt{3}}{3}$.

Example: Solve $16x^2 + 4x + 1 = 0$.

Solution: Move the 1 to the other side:

$$16x^2 + 4x = -1.$$

Factor 16 out on the left-hand side:

$$16\left(x^2 + \frac{1}{4}x\right) = -1.$$

Take half of $\frac{1}{4}$, which is $\frac{1}{2} \cdot \frac{1}{4} = \frac{1}{8}$, and then square it: $\left(\frac{1}{8}\right)^2 = \frac{1}{64}$.

Add this inside the parentheses:

$$16\left(x^2 + \frac{1}{4}x + \frac{1}{64}\right) = -1.$$

Now, we are really adding $16 \cdot \frac{1}{64}$ to both sides, which is $\frac{1}{4}$:

$$16\left(x^2+\frac{1}{4}x+\frac{1}{64}\right)=-1+\frac{1}{4}.$$

We can always multiply this back out to make sure we end up with the same equation that we started with:

$$16\left(x+\frac{1}{8}\right)^2=-\frac{4}{4}+\frac{1}{4}$$
$$16\left(x+\frac{1}{8}\right)^2=-\frac{3}{4}$$
$$\left(x+\frac{1}{8}\right)^2=-\frac{3}{64}$$
$$x+\frac{1}{8}=\pm\sqrt{-\frac{3}{64}}$$
$$x+\frac{1}{8}=\pm\frac{i\sqrt{3}}{\sqrt{64}}$$
$$x+\frac{1}{8}=\pm\frac{i\sqrt{3}}{8}$$
$$x=-\frac{1}{8}\pm\frac{i\sqrt{3}}{8}.$$

So

$$x=\frac{-1-i\sqrt{3}}{8} \quad or \quad x=\frac{-1+i\sqrt{3}}{8}.$$

Completing the square is helpful because it offers us a first look at being able to solve any quadratic equation, no matter what the coefficients are—and even if we cannot factor it.

Factors and Quadratics in Other Forms

Finding the equation that certain solutions come from is essentially a matter of working backward from what we have been doing.

Example: Find a quadratic equation that has solutions 2 and 6.

Solution: If $x = 2$ and $x = 6$, we can move the constants to the other side to say that

$$x - 2 = 0 \quad or \quad x - 6 = 0.$$

Then we can use the zero-product property in reverse to say that the product of these factors must also be zero. So we have

$$(x-2)(x-6) = 0.$$

FOIL this out, and we have our equation:

$$x^2 - 8x + 12 = 0.$$

Example: Solve $x^{-2} - 4x^{-1} - 5 = 0$.

Solution: This may not look like a quadratic equation, but suppose we substitute $y = x^{-1}$. Then, if we square this equation we get $y^2 = (x^{-1})^2 = x^{-2}$. So

$$x^{-2} - 4x^{-1} - 5 = 0$$

is equivalent to

$$y^2 - 4y - 5 = 0.$$

Now we can factor this equation to get

$$(y+1)(y-5) = 0.$$

By the zero-product property, $y = -1$ or $y = 5$. But we are solving for x, so we substitute x^{-1} back in as follows:

$$\begin{gathered} x^{-1} = -1, x^{-1} = 5 \\ \frac{1}{x} = -1, \frac{1}{x} = 5 \\ x = -1, x = \frac{1}{5}. \end{gathered}$$

When we plug both of these solutions back into the original equation, it checks out.

Example: Solve $y^4 - 8y^2 - 9 = 0$.

Solution: If we note that $y^4 = (y^2)^2$, then we can think of this as a quadratic equation in y^2, and we can factor this as we normally would, but with y^2 in our binomials. Alternatively, we can make the substitution $z = y^2$ and proceed as in the previous example. In that case, we would have

$$\begin{aligned} y^4 - 8y^2 - 9 &= 0 \\ (y^2 - 9)(y^2 + 1) &= 0 \\ y^2 - 9 = 0, \quad y^2 + 1 &= 0 \\ y^2 = 9, \quad y^2 &= -1 \\ y = \pm 3, \quad y &= \pm i. \end{aligned}$$

This equation has four solutions, $-3, 3, i, -i$.

YOU TRY IT: Check that all four solutions satisfy $y^4 - 8y^2 - 9 = 0$.

Section 8.2 Exercises

Solve the following equations.

1. $x^2 - 4x + 1 = 0$
2. $x^2 - 4x + 12 = 0$
3. $x^2 + 3x + 6 = 0$
4. $x^2 + 10x + 7 = 0$
5. $x^2 = -7 - 10x$
6. $3 = 8x - x^2$
7. $y^2 - 2y + 2 = 0$
8. $z^2 + z - 1 = 0$
9. $2x^2 + 4x + 5 = 0$
10. $3x + 1 = -6x^2$
11. $4x^2 - 2x + 5 = 0$
12. $10x^2 + 15x - 1 = 0$
13. $2x^2 + 2x - 1 = 0$
14. $8x^2 + 2x - 1 = 0$

15. Find a quadratic equation that has solutions 2 and –7.
16. Find a quadratic equation that has solutions 4 and –4.
17. Find a quadratic equation that has solutions $1+\sqrt{3}$ and $1-\sqrt{3}$.
18. Find a quadratic equation that has solutions $2+3\sqrt{2}$ and $2-3\sqrt{2}$.
19. Find a quadratic equation that has solutions $4i$ and $-4i$.
20. Find a quadratic equation that has solutions $5+12i$ and $5-12i$.
21. Solve $x^{-2}+x^{-1}-2=0$.
22. Solve $y^4+y^2-12=0$.
23. Solve $x^4-5x^2+4=0$.
24. Solve $z^{-2}-z^{-1}-30=0$.
25. Solve $y^4-9y^2=0$.
26. Solve $x^4+3x^2=0$.
27. Looking ahead to conic sections: Complete the square on the equation $y=x^2-4x+5$ by moving 5 to the other side. Leave your answer in the form $y-k=(x-h)^2$.

8.3: THE QUADRATIC FORMULA

Completing the square is great when the coefficients of the quadratic and linear term are reasonably factored. But completing the square on an equation like

$$7x^2+9x-1=0$$

would be challenging indeed, with the fractions that would be involved. So we use the quadratic formula instead.

If $ax^2+bx+c=0$, and $a\neq 0$, then

$$x=\frac{-b\pm\sqrt{b^2-4ac}}{2a}.$$

This gives us an algorithm for solving any quadratic equation using the coefficients.

Important: Remember to use parentheses when evaluating the quadratic formula, as we discussed in Section 1.6.

Example: Solve $3x^2 + 4x - 1 = 0$.

Solution: In this equation, $a = 3$, $b = 4$, and $c = -1$. So

$$\begin{aligned} x &= \frac{-(4) \pm \sqrt{(4)^2 - 4(3)(-1)}}{2(3)} \\ &= \frac{-4 \pm \sqrt{16 + 4(3)}}{6} \\ &= \frac{-4 \pm \sqrt{16+12}}{6} \\ &= \frac{-4 \pm \sqrt{28}}{6} \\ &= \frac{-4 \pm \sqrt{4 \cdot 7}}{6} \\ &= \frac{-4 \pm 2\sqrt{7}}{6}. \end{aligned}$$

In this last expression, everything in the numerator is divisible by 2, so we can factor out a 2:

$$\begin{aligned} x &= \frac{2\left(-2 \pm \sqrt{7}\right)}{6} \\ &= \frac{-2 \pm \sqrt{7}}{3}. \end{aligned}$$

Thus

$$x = \frac{-2 - \sqrt{7}}{3} \quad or \quad x = \frac{-2 + \sqrt{7}}{3}.$$

Example: Solve $8x^2 - 7x + 2 = 0$.

Solution: This equation is not easily factored, and completing the square would be difficult, so let's try using the quadratic formula. Here, $a = 8$, $b = -7$, $c = 1$. We have

$$\begin{aligned} x &= \frac{-(-7) \pm \sqrt{(-7)^2 - 4(8)(2)}}{2(8)} \\ &= \frac{7 \pm \sqrt{49 - 32 \cdot 2}}{16} \\ &= \frac{7 \pm \sqrt{49 - 64}}{16} \\ &= \frac{7 \pm \sqrt{-15}}{16} \\ &= \frac{7 \pm i\sqrt{15}}{16}, \end{aligned}$$

so our solutions are

$$x = \frac{7}{16} - \frac{i\sqrt{15}}{16}, \quad or \quad x = \frac{7}{16} + \frac{i\sqrt{15}}{16}.$$

The Discriminant

Recall Section 5.5, where we discussed factoring completely. In terms of quadratic expressions, up until now it may be difficult to tell whether the equation has real solutions, let alone rational solutions. For example, consider the quadratic equations

$$x^2 - 4x + 5 = 0 \qquad and \qquad x^2 + 4x - 5 = 0.$$

They are extremely similar, yet one of them has integer solutions and the other has nonreal solutions. Which equation is which?

To describe the roots of a quadratic, look at its discriminant. The discriminant of $ax^2 + bx + c = 0$ is

$$b^2 - 4ac.$$

The discriminant is what's under the radical in the quadratic equation; it is what we take the square root of in order to determine the solutions. So if the discriminant is negative, we will be taking the square root of a negative number, and the solutions will be nonreal.

The quadratic equation $ax^2 + bx + c = 0$ has the following:

- Two nonreal solutions if the discriminant $b^2 - 4ac$ is negative,
- One unique real solution if the discriminant $b^2 - 4ac$ is zero,
- Two solutions that are real and rational if $b^2 - 4ac$ if a positive perfect square, and
- Two solutions that are real and irrational if $b^2 - 4ac$ is a positive number that is not a perfect square.

Example: Determine the nature of the solutions to $2x^2 + 4x + 3 = 0$ without actually solving the equation.

Solution: If we calculate the discriminant, $a = 2$, $b = 4$, and $c = 3$. So

$$\begin{aligned} b^2 - 4ac &= (4)^2 - 4(2)(3) \\ &= 16 - 24 \\ &= -8. \end{aligned}$$

Since the discriminant is negative, the equation has two solutions that are nonreal numbers. Knowing this, we shouldn't even try to factor the equation. We would have to complete the square or use the quadratic formula.

Example: What kind of solutions does the equation $x^2 + x - 1 = 0$ have?

Solution: We calculate the discriminant with $a = 1$, $b = 1$, and $c = -1$, and we get

$$\begin{aligned} b^2 - 4ac &= (1)^2 - 4(1)(-1) \\ &= 1 - (-4) \\ &= 1 + 4 \\ &= 5. \end{aligned}$$

Remember that in the quadratic formula, we would be taking the square root of this number. So we would have a $\sqrt{5}$ in the answer, which means the solutions would be irrational and real.

Some students prefer to use the quadratic formula, because it can be applied to any quadratic. Instead of determining if we can factor, or instead of trying to complete the square and possibly getting fractions using brute force, we can bring out the heavy artillery and apply the formula that works every time (when it's used correctly).

However, a case should be made for remembering how to *complete the square* after you finish this course. Completing the square can be used in many applications to transform equations. We will use it again in Section 8.5 to help us graph quadratic equations. Also, it is where the quadratic formula comes from!

If we complete the square on $ax^2 + bx + c = 0$, we can divide both sides by a—because this is an equation (we don't have to factor it like an expression) and because $a \neq 0$:

$$\begin{aligned} x^2 + \frac{b}{a}x + \frac{c}{a} &= 0 \\ x^2 + \frac{b}{a}x &= -\frac{c}{a}. \end{aligned}$$

Taking half the linear coefficient gives $\frac{1}{2}\left(\frac{b}{a}\right) = \frac{b}{2a}$. Squaring this gives $\left(\frac{b}{2a}\right)^2 = \frac{b^2}{4a^2}$, which we add to both sides:

$$x^2 + \frac{b}{a}x + \frac{b^2}{4a^2} = -\frac{c}{a} + \frac{b^2}{4a^2}.$$

The left-hand side should be easy to factor. The right-hand side needs a common denominator of $4a^2$.

$$\begin{aligned} \left(x + \frac{b}{2a}\right)^2 &= -\frac{c}{a} \cdot \frac{4a}{4a} + \frac{b^2}{4a^2} \\ \left(x + \frac{b}{2a}\right)^2 &= -\frac{4ac}{4a^2} + \frac{b^2}{4a^2} \\ \left(x + \frac{b}{2a}\right)^2 &= \frac{b^2 - 4ac}{4a^2}. \end{aligned}$$

Take plus/minus the square root of both sides:

$$x+\frac{b}{2a}=\pm\frac{\sqrt{b^2-4ac}}{2a}$$

$$x=-\frac{b}{2a}\pm\frac{\sqrt{b^2-4ac}}{2a}$$

$$x=\frac{-b\pm\sqrt{b^2-4ac}}{2a}$$

This is the quadratic formula!

Section 8.3 Exercises

Solve the following equations, using any method.

1. $x^2+3x+4=0$
2. $x^2+3x-4=0$
3. $2x^2+5x+10=0$
4. $2x^2+3x-4=0$
5. $x^2-x-10=0$
6. $x^2+2x-5=0$
7. $3x^2+7x-1=0$
8. $2x^2+2x-3=0$
9. $2x^2+2x+3=0$
10. $x^2-5x+1=0$
11. $4x^2+5x-1=0$
12. $3x^2+2x-3=0$
13. $x^2+x-1=0$
14. $2x^2+2x-1=0$
15. $x^2+4x+4=0$
16. $3x^2-5x+9=0$
17. $2x^2+4x-1=0$
18. $3x^2+3x-1=0$

Determine the type of solutions the following equations have, without actually solving the equation.

19. $x^2 + 2x + 1 = 0$
20. $3x^2 - 4x + 1 = 0$
21. $\frac{1}{2}x^2 - 5x + 1 = 0$
22. $2x^2 + 3x + 8 = 0$
23. $\frac{1}{2}x^2 + \frac{1}{4}x - 1 = 0$
24. $0.1x^2 - 0.2x + 1 = 0$
25. $11x^2 + 4x - 2 = 0$
26. $3x^2 + 8x - 7 = 0$

8.4: FUNCTIONS

Functions are a critical topic in mathematics, and they are used in all scientific fields to describe how two sets relate to each other.

A function is a rule that assigns an element from one set (called the domain) to exactly one element of another set (called the codomain, or range).

We use the notation $f(x)$ to denote a "function f of the variable x." So

$$f(x) = x^2$$

is an example of a function that takes an element x to x^2.

Often, we replace $f(x)$ with y, because it is helpful to think of $f(x)$ as depending on the independent variable x. Thus, $f(x) = x + 1$ is a linear function that has the same graph as the line $y = x + 1$.

Evaluating functions at points of the domain is just like evaluating expressions in Section 1.6: Use parentheses! So when we want $f(2)$, everywhere we see x we replace it with the number 2 surrounded by parentheses. Think of it like a machine that takes in x and returns $f(x)$, the function value of x.

Example: For the function $f(x) = -x^2 - x + 1$, evaluate $f(1), f(2)$, and $f(-1)$.

Solution: If we want $f(1)$, then everywhere we see x in the function we replace it with 1:

$$\begin{aligned} f(1) &= -(1)^2 - (1) + 1 \\ &= -1 - 1 + 1 \\ &= -2 + 1 \\ &= -1. \end{aligned}$$

For $f(2)$, we replace x with 2:

$$\begin{aligned} f(2) &= -(2)^2 - 2 + 1 \\ &= -4 - 2 + 1 \\ &= -6 + 1 \\ &= -5. \end{aligned}$$

Now, for $f(-1)$, we replace x with -1.

$$\begin{aligned} f(-1) &= -(-1)^2 - (-1) + 1 \\ &= -1 + 1 + 1 \\ &= 0 + 1 \\ &= 1. \end{aligned}$$

Thus $f(1) = -1$ and $f(-1) = 1$.

Consider the function $g(x) = 2x + 1$. We can think of $g(x)$ as the y-value for a given x. If $x = 0$, then $g(0) = 2(0) + 1 = 1$. The y-coordinate corresponding to $x = 0$ is 1, and the point (0, 1) is a point on the curve. We recognize this as the y-intercept of the line. As we discussed in Section 3.1, this makes x the independent variable and $g(x)$ the dependent variable. If we have a linear function $f(x) = ax + b$, we can replace $f(x)$ with y and think of this as the line

$$y = ax + b.$$

If we take this a step further and think of the graph of what a function might look like, recall that a function is allowed only one y-value for every x. This means that on a graph, it will be a function if any vertical line hits the graph at most once. This is called the Vertical Line Test.

Example: Is the following the graph of a function?

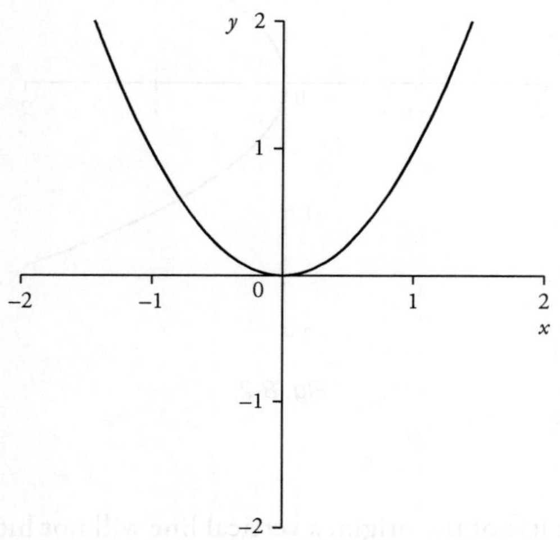

Fig. 8-1

Solution: If we draw vertical lines, they hit the function once. So Figure 8-1 is the graph of a function.

Example: Is the following the graph of a function?

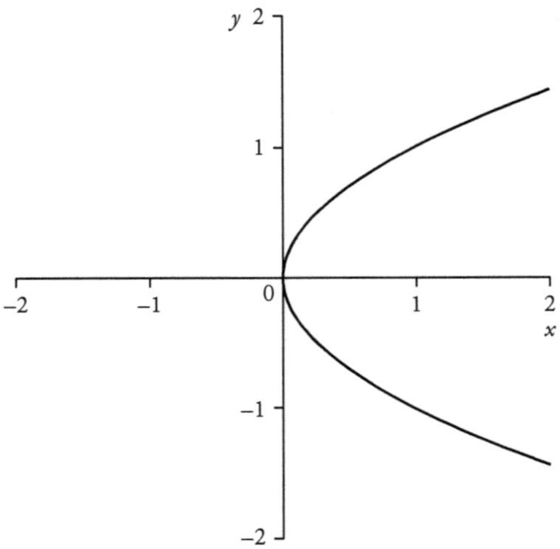

Fig. 8-2

Solution: Note that to the left of the origin, a vertical line will not hit the function at all. That's fine, but the problem lies to the right of $x = 0$. Any vertical line to the right of the origin will hit the graph twice. So the graph shown in Figure 8-2 is not the graph of a function.

Domain of a Function

The domain of a function is critical to our understanding of what x-values we can plug into the function.

When we try to find the domain of a function given a formula, it is easier to look at the values that we *cannot* plug in for x. These are the values that will be excluded from the domain.

Example: Find the domain of the function $f(x) = x^2 + 2x - 1$.

Solution: There is no number that we cannot plug in for x. We can take x and square it, multiply it by 2, add these together, and subtract 1. There are no problems, so we can say the domain is $\{x : x \in \mathbb{R}\}$, or we can simply say the domain is $\mathbb{R}$.

In fact, consider any polynomial $f(x)=a_nx^n+a_{n-1}x^{n-1}+\cdots+a_1x+a_0$. Since we can plug in any real number for x, *the domain of any polynomial is all real numbers.*

Ex) The domain of $h(x)=3x^{15}+8x^4-100x^3+30x^2-x+1$ is $\mathbb{R}$, because there is no domain restriction.

But this sweeping generalization does not apply to all functions. For the rational expressions and radical expressions that we've worked with, the domain is determined on a case-by-case basis.

Example: Find the domain of

$$f(x)=\frac{1}{x}+\frac{1}{x-1}.$$

Solution: We cannot allow division by zero, so 0 and 1 are not valid inputs for this function. However, anything else will be fine; any number besides 0 or 1 will return a function value. Recall from the set notation in Section 1.1 that we write the domain as

$$\{x\in\mathbb{R}: x\neq 0, 1\}$$

Example: Find the domain of the function $g(x)=\sqrt{x+1}$.

Solution: Another thing we have learned along the way is that if we take the square root of a negative number, the result will be nonreal. Most undergraduate mathematics courses deal with functions that have domains that are subsets of $\mathbb{R}$. So we cannot allow the square (or even) root of negative numbers. When finding the domain of a radical function like this, we need what's inside the radical to be nonnegative. In other words, it has to be at least zero. Better yet, we can use an inequality to say

$$x+1\geq 0.$$

So we see that $x\geq -1$ to get valid outputs. For example, $g(-1)=\sqrt{-1+1}=0$, and $g(3)=\sqrt{3+1}=\sqrt{4}=2$. But $g(-2)$ is not defined, because we do not allow outputs that are non-real. Outputs must be real. Therefore, our domain is

$$\{x\in\mathbb{R}: x\geq -1\}.$$

Example: Find the domain of $k(x)=\sqrt[3]{2-x}$.

Solution: For a given x, we would be calculating $2-x$ and then finding the cube root. Remember we can take the cube root of negative numbers. For example,

$$\begin{aligned} k(10) &= \sqrt[3]{2-10} \\ &= \sqrt[3]{-8} \\ &= -2. \end{aligned}$$

So for the cube root, there is no issue. The domain of $k(x)$ is all real numbers.

In general, we have to exclude negative numbers from radical functions only if there are *even roots* present. In order to obtain a real number, we need what's inside an even root to be nonnegative. But there are no restrictions on *odd roots*.

Different Ways to Represent a Function

A function is a rule, so a function does not always have to be defined with a formula. For example, we could say "the function that describes the rainfall in Orlando, Florida, over a 12-month period."

We can also use data to express a function.

Example: The following chart shows the average monthly rainfall in Orlando, Florida, in inches over a 12-month period. Does the data represent a function? What is the domain? What is the range?

Jan	Feb	Mar	Apr	May	June	July	Aug	Sept	Oct	Nov	Dec
2.76	2.83	3.78	2.48	3.31	8.74	7.09	7.83	6.02	3.31	2.4	2.64

(Source: usclimatedata.com)

Solution: The data represents a function, because each month has a corresponding rainfall amount in inches associated with it. (It would not be a function if we had two data values for a single month, which would not make sense.) Our domain is the set of months

{*January, February, March, April, May, June, July, August, September, October, November, December*}, and our range is the set of rainfall values

$$\{2.76, 2.83, 3.78, 2.48, 3.31, 8.74, 7.09, 7.83, 6.02, 3.31, 2.4, 2.64\}.$$

There are more abstract examples, where we can use sets. For example, suppose the domain for a function f is the set $D = \{1,3,5,6\}$ and the range is the set $R = \{2,5,8\}$. We can define the function so that

$$f(1) = 2$$
$$f(3) = 8$$
$$f(5) = 8$$
$$f(6) = 5.$$

We can represent this correspondence with ordered pairs (x, y), where the x-coordinates are the domain values and the y-values are the function values associated with x. So this function f can be described as the ordered pairs

$$(1, 2), (3, 8), (5, 8), (6, 5).$$

Section 8.4 Exercises

In the following exercises, compute $f(2)$, $f(1)$, and $f(-3)$.

1. $f(x) = x$
2. $f(x) = \sqrt{x+3}$
3. $f(x) = x^3$
4. $f(x) = -x^2$
5. $f(x) = 2x+3$
6. $f(x) = -x^2 + 3x$
7. $f(x) = x^3 + x^2$

For the following exercises, determine if the graph is of a function.

8.

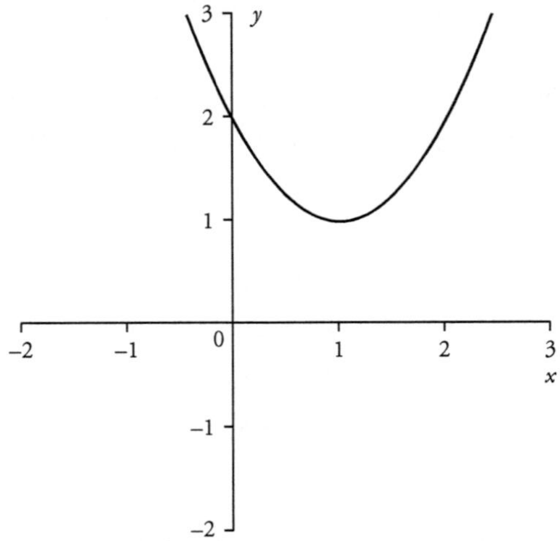

Fig. 8-3

9.

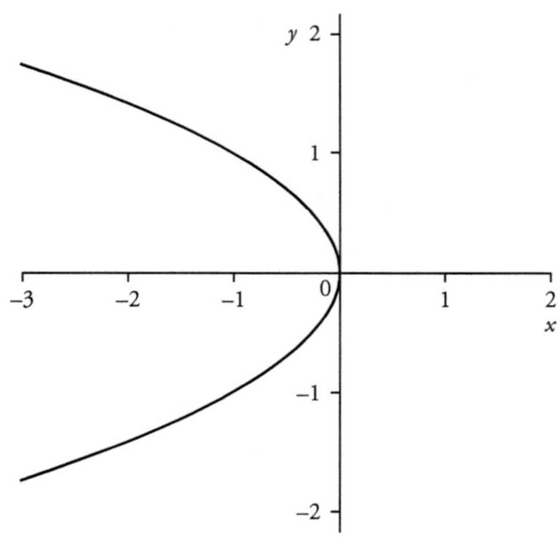

Fig. 8-4

10.

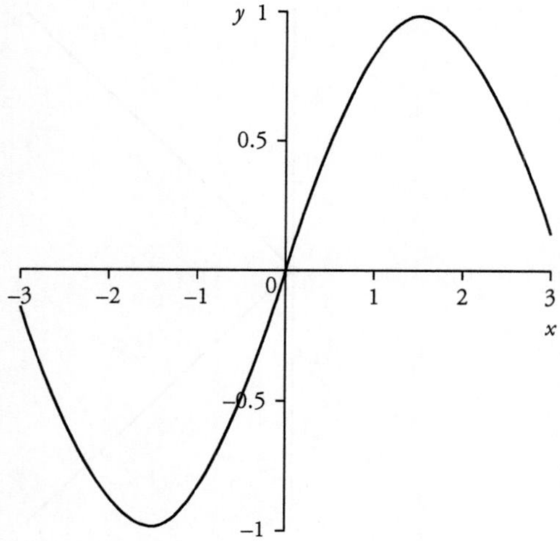

Fig. 8-5

11.

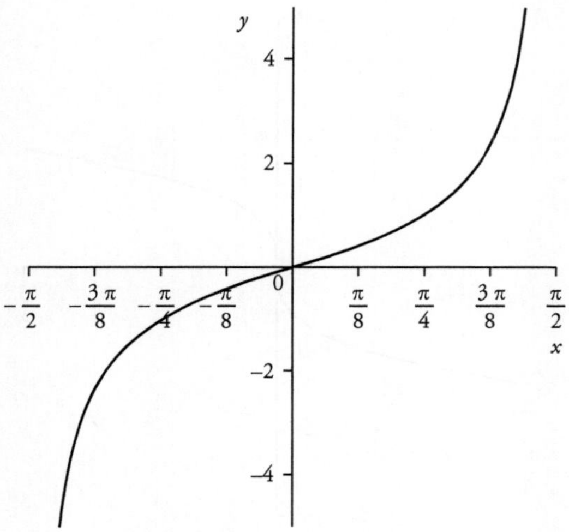

Fig. 8-6

12.

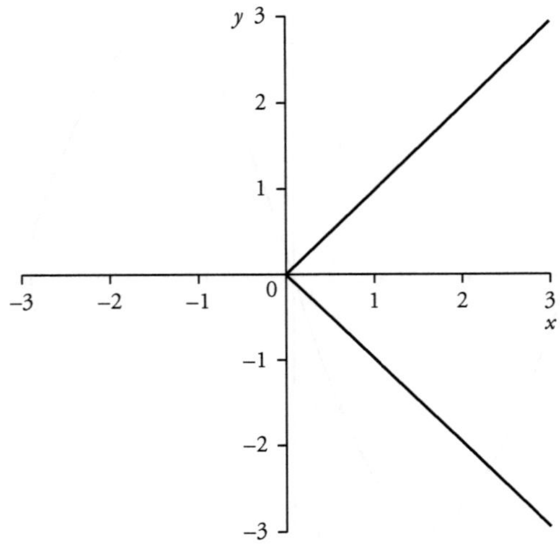

Fig. 8-7

13.

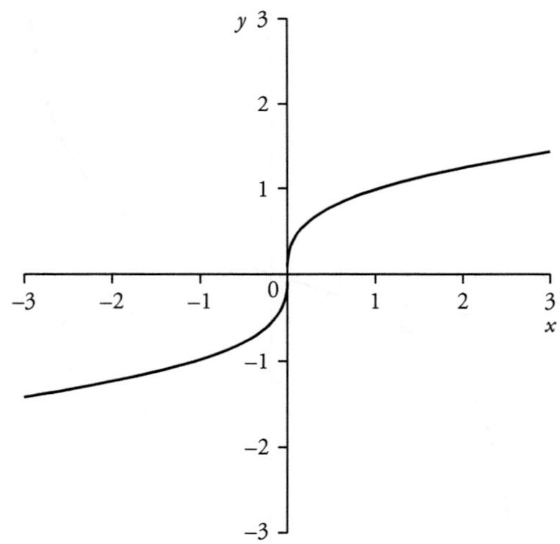

Fig. 8-8

Find the domain of the following functions.

14. $f(x)=\frac{1}{x}$

15. $g(x)=\frac{3}{x}$

16. $f(x)=x^2+x$

17. $g(x)=x^2+\frac{1}{x}$

18. $f(x)=\sqrt{x+2}$

19. $h(x)=\sqrt[3]{x+1}$

20. $f(x)=\frac{x+1}{(x-1)(x-2)}$

21. $f(x)=\frac{2x+1}{2x-1}$

22. $f(x)=x^2+\frac{2}{x-5}$

23. $f(x)=\sqrt{x-5}$

24. $f(x)=\sqrt{2x-5}$

25. $f(x)=x^{\frac{1}{3}}$

26. $f(x)=x^{\frac{1}{2}}$

27. $f(x)=x^{\frac{1}{3}}+x^{\frac{1}{2}}$

28. $f(x)=x^3+x^2$

29. The following gives the relationship between the amount of money an individual donated to a charity drive and the amount matched by a supporting business.
    - If an individual donated \$1, the company donated \$2.
    - If an individual donated \$3, the company donated \$8.
    - If an individual donated \$10, the company donated \$1,024.

    a. Is this relationship a function?
    b. What is the domain?
    c. What is the range?
30. As Emily was growing up, she kept track of her height in inches. Consider the data given below.

Year	1994	1996	1998	2000
Height (inches)	46	48	50	52

    a. Is this relationship a function?
    b. What is the domain?
    c. What is the range?
31. Consider the function $f(x) = 5 - 4x$.
    a. Give $f(0)$.
    b. Give $f\left(\frac{1}{4}\right)$.
    c. Give $f(-10)$.
    d. Give $f(a)$.
    e. Give $f(x+h)$.

8.5: Quadratic Functions

In this section, we will discuss how to graph a quadratic function

$$f(x) = ax^2 + bx + c.$$

Recall that a is the coefficient of the quadratic term, b is the coefficient of the linear term, and c is the constant.

A parabola has the property that the graph has one turning point to it. A parabola can open in any direction, but $f(x) = ax^2 + bx + c$ can open only upward or downward. The parabola is symmetric about its axis of symmetry, which is a vertical line that divides the parabola into two equal halves. At this point, we could fold the parabola over and it would lie on top of itself.

First, determine if the coefficient of the quadratic term is positive or negative. If it is positive, the parabola will open upward. If it is negative, the parabola opens downward.

Next, we want to locate the coordinates of the turning point of the graph. The way to do this will be to complete the square! If we complete the square on $f(x) = ax^2 + bx + c$, we will obtain the standard form of the equation:

$$f(x) = a(x - h)^2 + k.$$

At this time, we will complete the square only to transform the equation; we are not trying to solve for x yet.

The coordinates of the vertex are (h, k), and the axis of symmetry is the line $x = h$.

Example: For the graph of the quadratic function $f(x) = x^2 + 2x - 1$, give the coordinates of the vertex, determine the equation of the axis of symmetry, locate the x- and y-intercepts, and sketch its graph.

Solution: We begin by completing the square. First, we can move the constant to the other side. We have

$$\begin{aligned} f(x) + 1 &= x^2 + 2x \\ f(x) + 1 + 1 &= x^2 + 2x + 1 \\ f(x) + 2 &= (x + 1)^2 \\ f(x) &= (x + 1)^2 - 2. \end{aligned}$$

We have transformed the graph from general form into standard form. This should remind us of graphing lines. Just like when we were able to read the slope and y-intercept of a line directly from the equation in slope-intercept form, we are able to read the coordinates of the vertex from the quadratic function put in standard form.

Next, we take the opposite of what's with the x, which is –1. This is the x-coordinate of the vertex of the parabola. The y-coordinate is what's sitting by itself: the –2. So the coordinates of the vertex are $(-1,-2)$.

Note that $a=1$ here. In particular, a is positive. So the graph of the parabola opens upward.

To find the y-intercept for any function, we let $x=0$. We have $f(0)=-1$, so the y-intercept of the parabola is $(0,-1)$.

To find the x-intercept for any graph, we set $y=0$. Since $y=f(x)$, we set $f(x)=0$. By completing the square, we have already done a lot of the hard work. We have

$$\begin{aligned} 0 &= (x+1)^2-2 \\ 2 &= (x+1)^2 \\ \pm\sqrt{2} &= x+1 \\ -1\pm\sqrt{2} &= x, \end{aligned}$$

so the x-intercepts of the graph are $\left(-1-\sqrt{2},0\right)$ and $\left(-1+\sqrt{2},0\right)$.

A sketch of the graph is given in Figure 8-9.

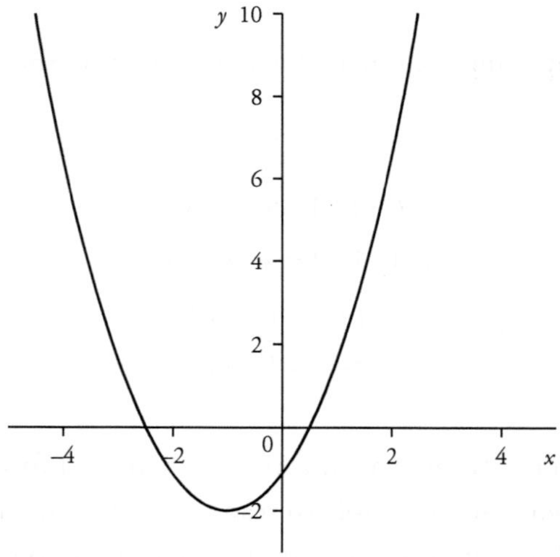

Fig. 8-9

Example: For $y = x^2 - 2x + 5$, determine the coordinates of the vertex, the x-intercepts (if any), the y-intercept, and the equation of the axis of symmetry. Then sketch the graph.

Solution: We complete the square first.

$$\begin{aligned} y &= x^2 - 2x + 5 \\ y - 5 &= x^2 - 2x \\ y - 5 + 1 &= x^2 - 2x + 1 \\ y - 4 &= (x-1)^2 \\ y &= (x-1)^2 + 4. \end{aligned}$$

We take the opposite of what is with the squared term, so the x-coordinate of the vertex is 1. The y-coordinate of the vertex is 4. So the vertex is located at (1, 4). Since the quadratic coefficient is positive, the parabola opens upward. To determine the x-intercepts, we set $y = 0$ and solve for x:

$$\begin{aligned} 0 &= (x-1)^2 + 4 \\ -4 &= (x-1)^2 \\ \pm\sqrt{-4} &= x - 1 \\ \pm 2i &= x - 1 \\ 1 \pm 2i &= x. \end{aligned}$$

Since these are not real numbers, there are no x-intercepts. Anytime an equation has nonreal solutions, there are no corresponding x-intercepts.

There is still a y-intercept. If we let $x = 0$, we have $y = 5$. So the parabola passes through $(0, 5)$. Finally, the axis of symmetry is the line $x = 1$. See the graph in Figure 8-10.

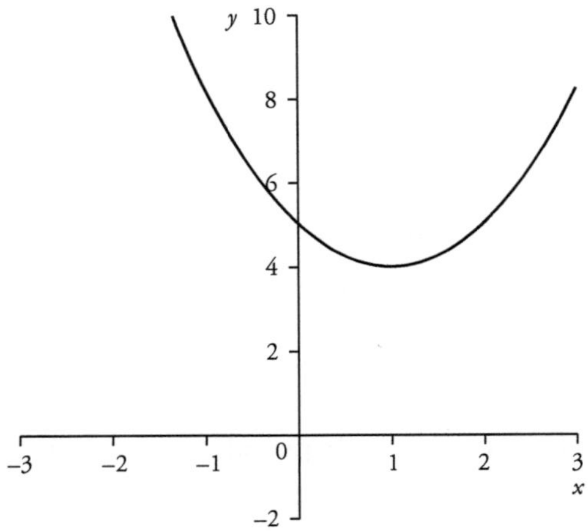

Fig. 8-10

Example: A tunnel is in the shape of a parabola. At its apex, it is 18 feet tall. The tunnel is 30 feet wide from one end to the other. Find an equation that models the tunnel.

Solution: A sketch of the parabola is given in Figure 8-11.

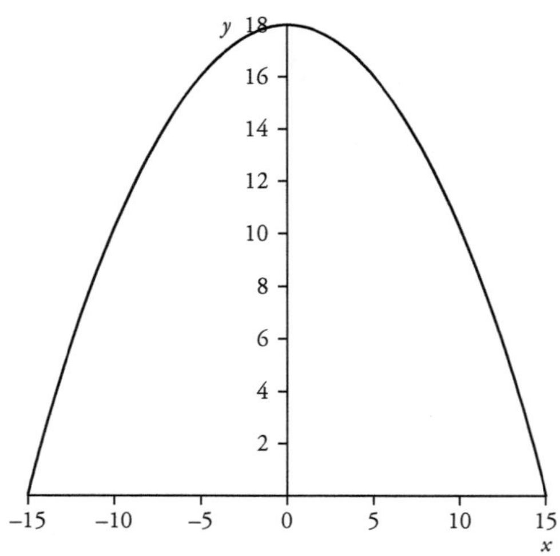

Fig. 8-11

Let's start with the standard equation $y = a(x-h)^2 + k$. For simplicity, let's make the axis of symmetry the y-axis. In other words, let $h = 0$. So we have

$$y = ax^2 + k.$$

We need to determine a and k. Now, the parabola is 18 feet tall, so the vertex of the parabola will be (0, 18). This means that $k = 18$:

$$y = ax^2 + 18$$

Finally, the tunnel is 30 feet wide from one end to the other. This means that both (15, 0) and (−15, 0) are x-intercepts for the equation. If we plug $x = 15$ and $y = 0$ into our equation, we have

$$\begin{aligned} 0 &= a(15)^2 + 18 \\ -18 &= 225a \\ -\frac{18}{225} &= a. \end{aligned}$$

We definitely want a to be negative, because the vertex is at (0, 18) and the parabola opens downward. An equation that models the tunnel is

$$y = -\frac{18}{225}x^2 + 18.$$

Section 8.5 Exercises

For the following equations, determine the coordinates of the vertex, the x-intercepts (if any), the y-intercept, and the equation of the axis of symmetry. Then sketch the graph.

1. $y = x^2$
2. $y = (x-1)^2$
3. $y = (x+1)^2$
4. $y = x^2 - 1$
5. $y = x^2 + 1$
6. $y = x^2 + 2x - 15$
7. $y = x^2 + 6x + 7$

8. $y = x^2 - 6x + 7$
9. $y = -x^2 - 2x - 2$
10. $y = -x^2 - 6x - 8$
11. $y = 4x^2 - 1$
12. $y = 3x^2 - x - 2$
13. $y = -3x^2 + 12$

14. A tunnel in the shape of a parabola is 40 feet wide at its widest point. If the tunnel is 20 feet tall at its highest point, write an equation that models the tunnel.
15. A boat passes through an archway that is 32 feet wide. The boat centers itself as it goes through the tunnel, and just makes it through. If the boat is 8 feet wide and 8 feet tall, write an equation that models the shape of the tunnel.
16. A two-dimensional sketch a parabolic satellite dish is 10 feet wide at the base, and it has a height of 5 feet. Determine an equation that models the drawing.
17. A marble is thrown from a height of 6 feet off the ground with an initial velocity of 8 feet/sec. If the equation of motion of the parabola is $y = -16t^2 + 8t + 50$, where t is time in seconds and y is given in feet, when does the marble hit the ground?
18. A marshmallow is dropped from a height of 100 m. If the height of the marshmallow can be modeled by $s(t) = -4.9t^2 + 100$, where t is in seconds and $s(t)$ is in meters, how long does it take for the marshmallow to reach the ground?

8.6: Quadratic Inequalities

In this last section, we will discuss quadratic inequalities. These are a step up from the linear inequalities that we discussed in Chapter 4. If $f(x)$ has the form $ax^2 + bx + c$, we are discussing inequalities of the form

$$f(x) > 0,\ f(x) < 0,\ f(x) \geq 0,\ f(x) \leq 0.$$

In general, the process outlined here for testing inequalities can be used for any function $f(x)$, not just for quadratics.

1. Solve the corresponding equation $f(x) = 0$.
2. Use these points to break up a real number line. If the inequality is strict ($<$ or $>$), the dots are not filled in. Otherwise ($\geq$ or $\leq$), the points are solid.

3. Pick points in the intervals created. Test these points in the original inequality. If this x satisfies the inequality, shade the interval this x came from. If it does not satisfy the inequality, do not shade it. Do this for each distinct interval.

Example: Graph the solution to the inequality $x^2 - 1 > 0$.

Solution: First we solve the corresponding equation $x^2 - 1 = 0$:

$$x^2 - 1 = 0$$
$$(x+1)(x-1) = 0$$
$$x+1=0 \quad or \quad x-1=0$$
$$x=-1, \quad 1$$

Now −1 and 1 are the testing points for our inequality. Since the inequality is strict, note that $x = 1$ and $x = -1$ do not satisfy $x^2 - 1 > 0$. So these will be open circles on our number line, and we draw a number line with −1 and 1 having open circles. See Figure 8-12.

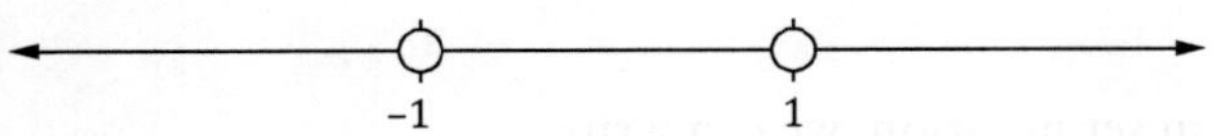

Fig. 8-12

Now we test the original inequality, which we can write factored as $(x+1)(x-1) > 0$.

We need a number less than −1, so let's try −5. When $x = -5$, we have

$$(-5+1)(-5-1) > ?0$$
$$(-4)(-6) > ?0$$
$$24 > 0.$$

So this works. On our number line, we can put a check mark above the interval less than −1 to denote that it works.

We need a number between -1 and 1, so it's easiest to test $x=0$. We have

$$\begin{aligned}(0+1)(0-1) &> ?0 \\ -1 &> ?0\end{aligned}$$

This is not true, so we put an X on the interval between -1 and 1.

Finally, we need a number greater than 1, so if we put in 100, we have

$$\begin{aligned}(100+1)(100-1) &> ?0 \\ 101\cdot 99 &> ?0\end{aligned}$$

Certainly this number is greater than zero, so any x-value larger than 1 should also work in the inequality. See Figure 8-13 for our completed test.

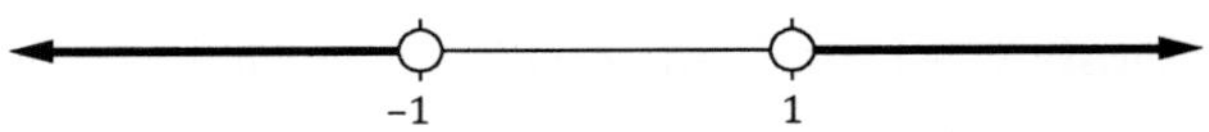

Fig. 8-13

To write our solution in set notation, we can write

$$\{x \in \mathbb{R} : x < -1 \text{ or } x > 1\}.$$

Example: Graph the solution to $8x^2 - 2x - 3 \le 0$.

Solution: First we need to solve $8x^2 - 2x - 3 = 0$. We can use the quadratic formula, but if we use $4x$ and $2x$ to begin to factor, we find

$$\begin{aligned}(2x+1)(4x-3) &= 0 \\ 2x+1 = 0 \quad &or \quad 4x-3=0 \\ x = -\frac{1}{2} \quad &or \quad x = \frac{3}{4}.\end{aligned}$$

With testing points $-\frac{1}{2}$ and $\frac{3}{4}$, we split up a number line with these two numbers. Since the inequality is a ≤, the holes at the test points are filled in. To test a number less than $-\frac{1}{2}$, let's choose −3. We test this in the original equality that has been factored:

$$(2x+1)(4x-3) \le 0.$$

Using $x = -3$, we test

$$\begin{aligned}(2(-3)+1)(4(-3)-3) &\le ?0\\ (-6+1)(-12-3) &\le ?0\\ (-5)(-15) &\le ?0.\end{aligned}$$

Since the left-hand side will be a positive number, we know that $x = -3$ does not satisfy the inequality. So we do not shade to the left of $-\frac{1}{2}$ on the number line.

To test a number between $-\frac{1}{2}$ and $\frac{3}{4}$, we can use 0:

$$\begin{aligned}(2(0)+1)(4(0)-3) &\le ?0\\ (1)(-3) &\le ?0\\ -3 &\le 0\end{aligned}$$

Since this inequality holds, we shade between $-\frac{1}{2}$ and $\frac{3}{4}$ on our number line.

Finally, we test larger than $\frac{3}{4}$, perhaps when $x = 1$:

$$\begin{aligned}(2(1)+1)(4(1)-3) &\le ?0\\ (3)(1) &\le ?0.\end{aligned}$$

Since this inequality is not true, we don't shade above $\frac{3}{4}$ on the number line. Our solution is graphed in Figure 8-14.

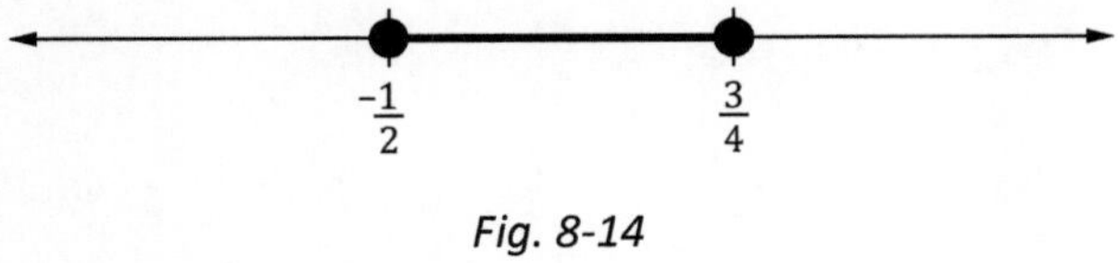

Fig. 8-14

If a quadratic equation cannot be factored easily, we can still use the same process: finding the roots by completing the square or using the quadratic formula.

Example: Graph the solution to $2x^2 - 11 < 0$.

Solution: Solving $2x^2 - 11 = 0$, we have

$$\begin{aligned} 2x^2 - 11 &= 0 \\ x^2 &= \frac{11}{2} \\ x &= \pm\sqrt{\frac{11}{2}}. \end{aligned}$$

Using a calculator gives us $x = \pm 2.345$. Testing a number less than -2.345, we test $x = -10$ in

$$2x^2 - 11 < 0.$$

We have

$$\begin{aligned} 2(-10)^2 - 11 &<? 0 \\ 2(100) - 11 &<? 0 \\ 200 - 11 &<? 0. \end{aligned}$$

This inequality is not true, so we don't shade below -2.345 on our number line. Testing $x = 0$, we have

$$\begin{aligned} 2(0)^2 - 11 &<? 0 \\ -11 &< 0. \end{aligned}$$

This inequality is true, so we shade between -2.345 and 2.345 on our number line. Finally, if we test $x = 10$, we see that $2 \cdot 10^2 - 11$ is certainly not less than zero, so we don't shade above 2.345 on our number line.

Since the inequality is strict, we have open circles at ±2.345. The solution is graphed in Figure 8-15.

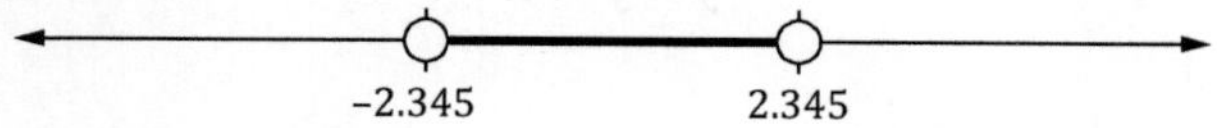

Fig. 8-15

Let's consider an example where there are no real solutions to the corresponding equation.

Example: What is the solution to $x^2+7\geq 0$?

Solution: Consider the equation $x^2+7=0$. We have

$$x^2=-7$$
$$x=\pm\sqrt{-7}$$
$$x=\pm i\sqrt{7}.$$

Since there are no real solutions to the equation, the corresponding inequality $x^2+7\geq 0$ will either always be true, or never be zero. There are no points that divide up the real line, so let's just test $x=0$. Since

$$0^2+7\geq 0,$$

any real number will solve this inequality. So the solution is $\mathbb{R}$.

As a final note, if we wanted to graph a quadratic inequality in two variables

$$y\geq ax^2+bx+c,$$

it would also be like when we graphed linear inequalities in Section 4.4. We would replace the inequality with an equals, and graph the parabola. If it is a strict inequality, we draw the parabola dashed. Otherwise the parabola is solid. Then if the inequality is

$$y>f(x)$$

or

$$y \geq f(x),$$

and we shade above the parabola. If the inequality is

$$y < f(x)$$

or

$$y \leq f(x),$$

then we shade below the parabola. Alternatively, we can test points as we did in Section 4.4.

Example: Graph the solution to $y \leq x^2 + 4$.

Solution: Consider the equation $y = x^2 + 4$. In the previous section, we discussed equations like $f(x) = a(x-h)^2 + k$. According to this, the vertex of $y = x^2 + 4$ should be (0, 4). The parabola opens upward since the coefficient of the quadratic term is positive here. And finally, we shade below the parabola since this is a "≤" inequality. See the solution below in Figure 8-16.

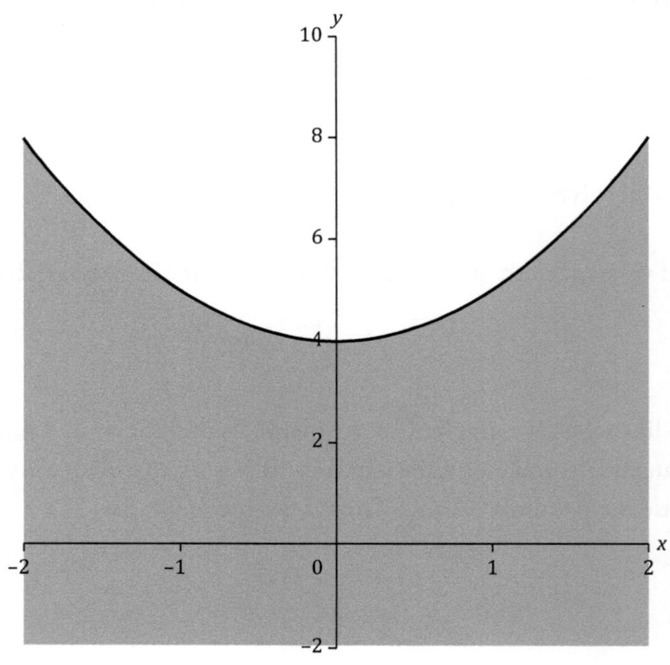

Fig. 8-16

Section 8.6 Exercises

Graph the solution to the following inequalities.

1. $x^2-4<0$
2. $9x^2-100\geq 0$
3. $x^2-16\leq 0$
4. $8x^2-25>0$
5. $x^2+5x-14\leq 0$
6. $x^2+6x+5\geq 0$
7. $(x+6)(x-2)<0$
8. $x^2+7x-44>0$
9. $x^2-10\geq 0$
10. $1-2x^2\geq 0$
11. $x^2>28$
12. $64<5x^2$
13. $x^2-4x-3>0$
14. $x^2+2x-4<0$
15. $x^2-14x+47\geq 0$
16. $x^2-6x+6\leq 0$
17. $x^2-2x+5<0$
18. $x^2-6x+25<0$
19. $x^2+9\leq 0.$
20. $x^2+9\geq 0.$
21. $x^2>-7.$
22. $2x^2+100<0$
23. $y>x^2$
24. $y\leq x^2$
25. $y\geq -x^2$
26. $y<-x^2$
27. $y>(x-1)^2+1$
28. $y\leq (x-1)^2+1$
29. Looking ahead: We can handle more factors in just the same way. For example, if we try to solve $(x+1)(x-3)(x-7)\leq 0$, we follow the same procedure as in this section. We set $(x+1)(x-3)(x-7)=0$. By the zero-product property, the three solutions are $-1,3,7$. Then we test the original inequality to see where the solution is.
    a. Graph the solution to $(x+1)(x-3)(x-7)\leq 0$.
    b. Graph the solution to $x(x-2)(x-4)\geq 0$.

Chapter 8 Review

8-A

1. Solve $x^2 - 25 = 0$.
2. Solve $x^2 + 4x + 3 = 0$ by factoring.
3. Solve $x^2 + 4x + 5 = 0$ by completing the square.
4. Solve $3x^2 + 5x - 1 = 0$ using the quadratic formula.
5. Solve $x - \frac{4}{x} = 0$.
6. Solve $\sqrt{x} - \sqrt{x-1} = 5$.
7. Solve $y^{-2} - 1 = 0$.
8. Find a quadratic equation that has solutions 2 and 3.
9. a. Find the domain of $f(x) = \frac{1}{x}$.

   b. Evaluate $f\left(\frac{1}{2}\right)$.
10. For $y = (x+1)^2 - 4$, determine the coordinates of the vertex, the x-intercepts (if any), the y-intercept, and the equation of the axis of symmetry. Then sketch the graph.
11. Graph the solution to the inequality $x^2 - 3x - 4 > 0$ on a number line.

8-B

In questions 1–7, solve the equation using any method.

1. $x + 1 = \frac{30}{x}$
2. $2x^2 - 7x + 1 = 0$
3. $4x^2 - 8x + 7 = 0$
4. $\sqrt{2x+5} = \sqrt{x+3} - \sqrt{x-1}$
5. $x^2 + 12x + 35 = 0$
6. $y^4 - 5y^2 - 36 = 0$
7. $2x^2 = 5x$

8. Find a quadratic equation with solutions $2+2\sqrt{3}$ and $2-2\sqrt{3}$.
9. a. If $g(x)=\sqrt{3x-5}$, give the domain of g.

   b. Evaluate $g\left(\frac{14}{3}\right)$.
10. For $y=-x^2-2x-5$, determine the coordinates of the vertex, the x-intercepts (if any), the y-intercept, and the equation of the axis of symmetry. Then sketch the graph.
11. Graph the solution to the inequality $6x^2-7x-5<0$ on a number line.

Printed by Libri Plureos GmbH in Hamburg,
Germany

9 781516 503032